Gustav von Wex

First and Second Treatise on the Decrease of Water in

Springs, Creeks, and Rivers

Contemporaneously with an Increase in Height of Floods in Cultivated

Countries

Gustav von Wex

First and Second Treatise on the Decrease of Water in Springs, Creeks, and Rivers
Contemporaneously with an Increase in Height of Floods in Cultivated Countries

ISBN/EAN: 9783337393694

Printed in Europe, USA, Canada, Australia, Japan

Cover: Foto ©berggeist007 / pixelio.de

More available books at **www.hansebooks.com**

FIRST TREATISE

ON THE

DECREASE OF WATER

IN

SPRINGS, CREEKS, AND RIVERS,

CONTEMPORANEOUSLY WITH

AN INCREASE IN HEIGHT OF FLOODS IN CULTIVATED COUNTRIES,

BY

GUSTAV WEX,

IMPERIAL AND ROYAL MINISTERIAL COUNSELLOR AND CHIEF ENGINEER
OF THE IMPROVEMENT OF THE DANUBE, AT VIENNA,

WITH SEVEN SHEETS OF DRAWINGS.

FROM THE PAPERS OF THE SOCIETY OF AUSTRIAN ENGINEERS AND
ARCHITECTS, 1873—Nos. 2, 4, AND 6.

TRANSLATED BY

G. WEITZEL,

Major of Engineers, Brevet Maj. Gen. U. S. A.

WASHINGTON:
GOVERNMENT PRINTING OFFICE.
1881.

GENERAL: I have the honor to transmit to-day, by registered mail, the manuscript of my translation of the hydraulic work entitled "Ueber die Wasserabnahme in den Quellen, Flüssen und Strömen bei gleichzeitiger Steigerung der Hochwässer in den Culturländern," (On the decrease of water in springs, rivers, and streams contemporaneously with an increase in height of floods in cultivated countries,) by Gustav Wex, imperial and royal ministerial counsellor and chief engineer of the improvement of the Danube at Vienna.

This work is known as the first treatise of the author on this subject. Since its publication in 1873, he has been knighted. I translated his second treatise last year at your request, and it was published at the Government Printing Office. A copy of this translation came into the hands of Sir Gustav Wex, whereupon he wrote me a very kind and courteous letter of thanks for the trouble which the translation caused me, and at the same time expressed his high appreciation of the honor which had been conferred on him by our Government in publishing the translation. At the same time he sent me a copy of his first treatise. I was informed by you that if I would translate this also, my translation would be published.

This work is especially important, since it not only completes the former translation, but contains very interesting information concerning the improvement of the Elbe River, which I doubt if many engineers have heretofore been able to obtain.

Very respectfully, your obedient servant,

G. WEITZEL,
Major of Engineers.

Brig. Gen. H. G. WRIGHT,
Chief of Engineers, U. S. A. Washington, D. C.

[INDORSEMENT.]

OFFICE OF THE CHIEF OF ENGINEERS,
UNITED STATES ARMY,
October 24, 1881.

Respectfully submitted to the Honorable the Secretary of War, recommending that the translation be printed at the Government Printing Office, and that 500 copies be furnished on requisition from this Office.

H. G. WRIGHT,
Chief of Engineers,
Brig. & Bvt. Maj. Gen.

Approved:
By order of the Secretary of War.

JOHN TWEEDALE,
Acting Chief Clerk.

OCTOBER 25, 1881.

3

INTRODUCTION.

It is universally known that our earth passed through powerful revolutions and transformations during its earlier periods, and there can be no doubt that, even at present, important changes are taking place on its surface. These are, however, no longer on so grand a scale nor so powerful, and even if they are constant, mostly take place so gradually that they are scarcely perceptible to the living generation, and can generally only be observed and established by observers who are supplied with scientific means of assistance.

A combination of kindred observations, a comparison of observations and plotting of the stages of rivers during periods of many years, and the results of thorough study which I made upon this subject, furnished me with many proofs of the continued decrease in the discharge of springs as well as in the creeks, rivers, and streams in the greater part of Europe, in those countries which have been cultivated during the historic period, and particularly those which are now or have been recently under cultivation.

It must be clear to all that such a continual decrease in the water supply must be considered as a very important hydrographic change of the earth surface.

If it is constantly kept before our eyes that neither vegetable nor animal life can exist without gathering water from its immediate neighborhood; that the richest lands owe their wealth and consequent ability to sustain a numerous population to this benevolent element, and that man, among all living beings, consumes the most fresh water in all possible forms and employs it for the greatest variety of purposes, it will be seen at once that the question here is not to discuss a phenomenon which is alone interesting to a friend of natural sciences, but to establish a fact of the highest practical importance. It concerns furthermore the investigation of the causes and effects of an evil which, if it gains the upper hand, will place in question the existence of future generations in the most thickly settled and richest in cultivation of all the abodes of man. The once rank and fruitful lands of Persia, Palestine, Greece, Sicily, and Spain, which are already partially arid and waste, bring before our eyes the saddest illustration of this.

If this continual decrease of the quantity of water which is poured upon and flows on the surface of the earth, and which has been established as a fact by us, were the result of natural forces at work in an unknown manner, there would remain nothing but to allow blind fate to reign and to consign future generations to their inevitable doom. This is, however, fortunately not the case. I believe that I may positively assert, as the result of my investigations, that the above-mentioned phenomenon in cultivated countries is mainly if not entirely due to events which are recorded in the history of culture. These are the effects of the works of man on the aspect of that portion of the earth's surface inhabited by him which, although apparently quite trivial, must not be underrated. I refer particularly to the selfish and thoughtless manner of reaping the products of the soil and the improper change in

5

the form of the surface which has thus been robbed. It is therefore possible to prevent, or at least to prolong, the threatening calamity through an incalculable period of time by a rational effort and by a unanimous, tireless resistance on a suitably great scale.

It is the duty of living generations, after being convinced of the presence of this evil and the possibility of its cure, to take hold of this work and mark out the path for its improvement for the next genera-tion. I can assume this to be the opinion of every one who recognizes in himself a member of the human family; who does not wish to repay with base selfishness toward his successors the great multitude of good things for which his gratitude is due to his ancestors, and one who is fully impressed with the truth of the maxim that states are instituted and created to secure the welfare of the whole of its members for all time.

This treatise is therefore particularly addressed to the higher govern-ments, and particularly to those branches of them to which are intrusted the care of the national economical interests of its inhabitants. For it lies in their province, and it is their duty, to cause such measures to be adopted and placed into execution as the magnitude of the danger requires. They alone can undertake the work with prospect of success by keeping the distant end steadily in view, and by means of the power of the government pressing measures vigorously, which should be of sufficient magnitude and harmonize clashing private interests.

I feel too that I should, before all, submit these results of my studies to the examination and criticism of my esteemed colleagues and gen-erally to all friends of the natural sciences. For my single note of warning would be like a voice in the desert if the facts and conclusions which led to this phenomenon were not thoroughly examined and found correct by competent authorities from as many different places as possible, and these did not warmly recommend to their higher govern-ments the execution of the measures and means which are recommended to prevent the continuance of the evil.

I desire particularly a thorough criticism of this treatise by my esteemed colleagues, for the reason that engineers particularly come more into contact with large owners of land and forests, large manu-facturers, and generally with persons who devote, in their own interest, a greater attention to national economical questions like this, and have more occasion and are better qualified to judge of them. This is par-ticularly true of hydraulic engineers, who are so situated that they can give my theory practical effect by giving due consideration in their projects for and execution of river improvements, aqueducts, and other hydraulic works, to the continued decrease in the discharge of flowing water. If they do not, their works will after a few decades no longer be suitable for the purpose for which they were built, or will be consid-ered as failures, as I will show to have been the case in several large hydraulic constructions.

In this treatise I have presented the proofs of the decrease in the height of stages and discharge of streams in such a detailed and variegated manner for the only reason that, in verbal and written con-ferences on this subject with colleagues and several naturalists, I have found several who oppose the conclusions which I have drawn from my researches. I therefore found myself induced, in this treatise, to men-tion and combat the objections and counter-arguments of these oppo-nents.

In conclusion, I take the liberty of remarking that I ought really, as bearing upon the natural formation and flow of waters, at first produce proofs of the decrease in the supply of water flowing from springs. There have up to the present time, however, been only a few reliable measurements made of the discharge of springs, and no reliable conclusions could be drawn relative to the conditions of discharge of the springs which number millions, even if these measurements had been made at hundreds of them. I will therefore furnish the data relative to the conditions of discharge of the larger streams of Europe, and from these can be deduced the discharge of the springs in their basins as a whole and in a reliable manner.

I. DECREASE IN THE HEIGHT OF STAGES IN RIVERS AND STREAMS AND THE RESULTING DECREASE IN THEIR DISCHARGE.

While engaged in preparing projects for the improvement of several of the larger rivers and streams situated in nearly all the crown lands of the monarchy, to which duty I was assigned by the Imperial and Royal Austrian Government, I first studied the relative discharges of these rivers, and found in almost every case a considerable decrease in the heights of the stages of these water-courses from those that had been observed many years ago.

This fact had already been observed by several naturalists, and they had based the opinion on it that the discharge of many rivers and streams *seemed* to decrease. The correctness of this opinion was, however, doubted and combated by many other naturalists, and particularly by several hydraulic engineers. I will therefore first enumerate the views of the latter, and afterwards oppose them by tabulated observations taken on the chief streams of central Europe during periods of many years.

Mr. Hagen, the royal Prussian chief privy counsellor and chief director of public works, in the third edition of his hand book on hydraulic art, published in 1871, doubts the correctness of the gauge-readings heretofore made on rivers and streams, and the conclusions drawn therefrom relative to the decrease in their discharge, and bases his opinion to the contrary on the recorded gauge-readings taken on the Rhine at Düsseldorf since 1800, a period therefore of 71 years.

Mr. Hagen calculated the arithmetical mean of the daily gauge-readings for each year, that is, the mean annual gauge-reading, and plotted those (but unfortunately on a very small scale) together with the highest and lowest gauge-readings of the year, from which he asserts that it follows that a considerable decrease in discharge has not taken place.

On account of the great importance of the question, Mr. Hagen subjected these gauge-readings to a calculation by the method of least squares, and found that during the period of observation on the Rhine at Düsseldorf, of 71 years, the equally inclined average lines sought for by Mr. Hagen show a depression of 2.9 lines for floods; 1.6 lines for the mean stages, and an elevation of those for low water of 0.2 lines annually. But Mr. Hagen himself attaches no value to this, since the possible error of these results amount to 2.2, 0.9, and 0.7 lines, respectively.

Mr. Hagen further says that it may be supposed that in the course of time a decrease in the height of floods and mean stages have occurred, and for the reason that the improvements of the rivers which have been made in late years have prevented the formation of ice-gorges and accelerated the discharge of floods, and that this explains the small annual decrease in the height of mean stages.

After Mr. Hagen had also declared that the large decrease in the height of water stages and in the discharge of the Weser, which several hydraulic engineers had stated as a fact from a comparison of the gauge-readings at Minden and Schlüsselburg, was improbable, he arrives at the following conclusion, viz:

So far as the observations which have been made admit of the formation of any opinion, the result does not show any general decrease in the discharge of the Weser or the Rhine.

Although this opinion, which Mr. Hagen expresses from the readings of these gauges on two streams, against the decrease in their discharge is quite reserved, yet most hydraulic engineers have come to the same conclusion in regard to other streams, because they did not have the gauge-readings taken during many years at their command and thus were unable to compare them, and then again because they consider Mr. Hagen an infallible authority.

Mr. Maass, royal Prussian inspector of hydraulic works, published the gauge-readings at Magdeburg, on the Elbe, for the period of 143 years, from 1727 to 1869, together with a plot thereof, in the Journal of Architecture by Erbkam, 1870, and thereby furnished the proof that during that period the arithmetical mean of the Elbe decreased for.—

	Feet.	Inches.
Floods	1	5.64
Lowest stages	2	11.27
And mean annual stages	3	0.83

Mr. Maass does not consider that this decrease in the height of the stages or water-surface is the result of a decrease in the discharge of the streams, but that it is due to the improvement of the river and the consequent deepening of its bed and increase of velocity. Mr. Maass silences the apprehensions which have been created by the sinking of the water surface, and which became so plainly visible, by stating that a further sinking in the same ratio need not to be feared, as no further works of improvement would be executed on the Elbe.

As these obervations and views of Messrs. Hagen and Maass are contrary to the observations which I have made, and the views which I have formed during an experience of forty years on many rivers, I resolved to tabulate the data and observations taken on several rivers and streams which are necessary in discussing this important question, in order to obtain in this manner perfectly reliable figures, and without which I could not venture successfully to antagonize the views of Mr. Hagen, who in all questions relating to hydraulics is the acknowledged authority throughout Germany.

In prosecuting my researches on this subject I came across that excellent work "Allgemeine Länder und Völkerkunde von Dr. Heinrich Berghaus, zweiter Band, Umrisse der Hydrographie vom Jahre 1837,"* and I found in this as well as in the work from the same author, but published by J. Perthes, entitled † "Hydro-historishen Uebersichten der deutschen Ströme vom Jahre 1838," quite valuable material on the hydraulic conditions and ice-flows of streams already grouped. This distinguished hydrographer, Mr. Berghaus, compared the highest and lowest and the annual mean of gauge-readings and plotted some of them, which were taken at Emmerich (at the Dutch frontier) on the Rhine during the period of 66 years from 1770 to 1835; at Cologne during that of 54 years from 1782 to 1835; Magdeburg on the Elbe during that of 10 years from 1728 to 1835; and at Küstrin on the Oder that of 58

* General Knowledge of Countries and People, second volume: Outline of Hydrography, 1837.

† Hydro-historical review of German streams, 1838.

years from 1778 to 1835. He then depicts in so detailed and thorough manner, as is not done in any other work on hydraulics, all the phenomena which result from these readings and which relate to the actual condition of the stages of water in the various months and years; the advent of ice-flows and floods, and finally the effect of atmospheric precipitations on the discharge of streams.

In this work Berghaus has at the same time pointed out the manner in which gauge-readings on rivers and streams should be grouped and compared in order to obtain a clear presentation of the hydraulic phenomena or life of a stream, and in order that correct conclusions may be drawn therefrom relative to its discharge and the changes therein.

I will now point out the chief results which are obtained from the very detailed and highly interesting data furnished by Berghaus.

On Sheet 1 I have plotted the observed highest and lowest and the calculated annual mean of gauge-readings taken during the period from 1770 to 1835, as given in the tables contained in Dr. H. Berghaus' hydrographic work. From this plot it will be seen that the water during individual years falls quite irregularly, rises and then falls again, and that therefore it is very difficult to calculate reliably an equally inclined line to represent the fall or rise of the mean heights during a long period, as Mr. Hagen himself admits. I therefore consider the method which Mr. Berghaus has adopted, of comparing the arithmetical mean of the readings during two long periods of time with each other in order to ascertain if the stages of the river have increased or diminished in height as the most reliable and simplest. If now the period of observation of 66 years is divided into two periods of 33 years, (from 1770 to 1802 and 1803 to 1835,) and we calculate the mean of the guage-readings of these periods, we will find that the mean height of floods during the latter period has increased 10 lines; the annual mean has decreased 1 foot 5 inches, and the lowest stages had decreased 1 foot 2 inches from those of the former period between 1770 and 1802.

Since the arithmetical mean of the guage-readings during the first period, from 1770 to 1802, were only taken for the sake of comparison and without regard to whether during these 33 years an increase or decrease had already been taking place, it is clear that the above figures only give the decrease in the mean and lowest stages during the several periods of 33 years, that is, for one-half the period of observation. The latter must be borne in mind in future in comparing the gauge-readings of several periods.

If again we divide the period of observations from 1782 to 1835, during which gauge-readings were taken at Cologne on the Rhine, and which are given in the tables at the end of Berghaus' book, into two periods of 27 years each, and calculate the means, we will find that from 1809 to 1835 the mean of flood stages has increased 1½ inches; the mean of annual stages has decreased 4½ inches; and the mean of the lowest stages 7½ inches compared with the former period from 1782 to 1808.

From this it will be seen that the two chief gauge stations on the Rhine, that is, at Emmerich during the period of 66 years from 1770 to 1835, and at Cologne during the period of 54 years from 1782 to 1835, show that a decrease which is not inconsiderable has taken place in the annual mean and mean of lowest stages, and a slight increase in the mean of flood stages in the older from the later half of the periods.

Since no works of improvement had been undertaken on the Rhine between 1770 and 1835 which could have caused at Cologne and Emmerich a deepening of the bed or an increase in the velocity of the current and a consequent depression of the water surface, the decrease in the height of the annual mean and low-water stages during the later periods can only be explained by a decrease in the discharge of the Rhine, and, on the other hand, a slight increase in the mean of flood stages must be ascribed to higher and more frequent rises.

The circumstance that the decrease in the height of the low and annual mean stages of the Rhine is less than that of other streams is explained by Berghaus as being the result of the Rhine being largely fed by the never-failing springs created by the masses of ice and snow in the Alps. I deem it proper that I should add to this the remark that the Lake of Constance, which collects the floods of the upper Rhine and permits them to flow off only gradually, regulates the conditions of discharge of the lower part of this stream to a considerable extent, and that therefore the decrease of discharge at the lower and mean stages can only show itself in a diminished degree.

Mr. Berghaus explained the circumstance that the decrease in the low and mean stages at Emmerich are greater than at Cologne by the fact that between Cologne and Emmerich the Rhine receives the tributaries Wupper, Ruhr, Emtsche, and Lippe, and these furnish a more irregular supply of water.

Now let us subject the plot of the gauge-readings at Düsseldorf on the Rhine, (that is, between Cologne and Emmerich,) taken during the period from 1800 to 1871, as published by Mr. Hagen, to a thorough examination.

Mr. Hagen found, according to his method, that the annual mean of the readings at Düsseldorf decreased on an average 1.6 lines per year, which would amount in fifty years to a decrease of nearly 7 inches.

Since, according to the tables of gauge-readings given by Berghaus, the decrease in the annual means for a period of 50 years would amount at Emmerich to 25¾ inches, and at Cologne to 8¼ inches, it will be seen that the decrease in annual means, determined by Mr. Hagen by calculation at Düsseldorf as nearly 7 inches, very nearly agrees with that determined at Cologne, although in a different period. This decrease is therefore a fact and not an error of calculation, as Mr. Hagen has said it probably is.

, Although according to the older tables of gauge-readings furnished by Berghaus, that is, between 1770 and 1835, there was an increase in the height of floods of from 10 to 18 lines, the decrease in the same calculated by Mr. Hagen to be on an average 2.9 lines per annum or 12 inches, 7 lines for 50 years would still be possible, because the velocity of discharge of the Rhine has been considerably accelerated since 1830 by the works of improvement which have been made between Basle and Mannheim.

The extremely slight increase in the height of low water at Düsseldorf, found by Mr. Hagen to average one-fifth of a line per annum, or only 10 lines per 50 years, is clearly explained by the circumstance that the bed has been somewhat narrowed there during past decades, and that in consequence of the cut-offs made in the Rhine improvements above Mannheim, sand and fine gravel were unavoidably carried into the lower river, and that thereby the bed at Düsseldorf was slightly filled and raised, as I will hereafter show was the case in the middle and lower Elbe in a still greater degree.

It will be seen from the foregoing how erroneous the conclusions of Mr. Hagen are that the observations which have generally been made on the Rhine do not warrant the belief that its discharge has decreased. If the other table of gauge-readings at Emmerich and Cologne, published by Berghaus, had been known to Mr. Hagen, and if he had more correctly compared those taken at Düsseldorf during a period of 71 years, he would have been convinced to the contrary.

In order to learn the conditions of discharge of the Rhine above Düsseldorf and Cologne, I turned for information on this subject to that distinguished hydraulic engineer, Grebenau,* royal Bavarian inspector of public works at Germersheim, who had been acquainted with the works of improvement on the Rhine during many years. I received from him accurate gauge-readings at Sonderheim, taken during a period of 28 years, from 1840 to 1867, as well as a statement of the annual discharge of the Rhine taken at the cut-off at Germersheim. Mr. Grebenau obtained these as accurately as possible by the exercise of great circumspection and professional knowledge, and after taking a great deal of pains and time, not only in making numerous direct measurements of the cross-section of the bed of the stream and velocity of discharge, but also by means of calculations made according to the newest formulæ and co-efficients determined by experiments.

If the arithmetical mean of the annual gauge-readings and annual discharge for the two periods of 14 years, into which the period of observation may be divided, is obtained from the plot on Sheet 2, it will be found that during the last half period the annual mean of gauge-readings has decreased 17½ inches, and the annual discharge about 7,227 cubic feet per second.

But it must be admitted that by far the greater part of this decrease in the annual mean is due to the depression of the water-surface, caused by the deepening of the bed of the stream, which was effected by means of the cut-offs which were made in that vicinity for the Rhine improvement.

And although we cannot be positively assured that the measurements and calculations made by Mr. Grebenau to determine this discharge during a period of 28 years are absolutely accurate and correct, and cannot deny that the gradual sinking of the bed produced by the river improvement had a great influence in decreasing the gauge-readings, yet being obtained by the same method, these figures giving the discharge must be nearly relatively correct, and we are fully justified by them in drawing the conclusion that at Germersheim also the discharge of the Rhine has decreased.

Then there is the striking phenomena that this beautifully improved, fairly canalized Rhine, between Mannheim and Basle, is no longer suitable as a water route in consequence of the numerous gravel bars formed near the concave banks, and which are continually moving down stream: of the continued, sharp, serpentining or winding of the channel, and the shallow crossings from one concave bend to the other. This can only be explained by the circumstance that the normal width of the improved stream was taken too great at the beginning, and that therefore the stream, in consequence of the decrease in its discharge, has no longer the power to move the gravel bars entirely and rapidly, but only sufficient to force them into the concave bends.

These results, obtained from observing four gauges on the Rhine, seem to me to be amply sufficient to corroborate my assertion in regard to this steam, and I therefore pass on to the data collected for other streams of Central Europe.

---

* Mr. Grebenau, who translated from English into German that great work of Humphreys and Abbot, entitled "Theory on the flow of Water in various Rivers and Canals," and added to it some water measurements of his own, was, in recognition of his prominent knowledge of hydraulics, appointed director of hydraulic works at Strassburg in 1871 by the Emperor of Germany.

Mr. Berghaus in his above-mentioned works has published tables of gauge-readings taken at Magdeburg, on the Elbe, from 1728 to 1835, and calculated from these the tables below. This table includes the full century from 1731 to 1830, and I invite special attention here to the last column. It will be seen from this that the mean monthly gauge-readings taken from 1781 to 1830 have decreased from 15.56 inches to 26.79 inches, and the annual mean 21.39 inches, from those taken during the half century from 1731 to 1780.

Table of decrease in the water stages of the Elbe at Magdeburg for separate months.

Months.	Mean stage during—		Decrease of stages in the second half of the century.
	First half century, from 1731 to 1780.	Second half century, from 1781 to 1830.	
	Ft. in.	Ft. in.	Ft. in.
January	9 8.48	7 11.14	1 9.54
February	10 6.61	8 11.63	1 6.98
March	11 6.59	10 2.79	1 3.80
April	11 4.47	10 0.67	1 3.80
May	9 9.19	8 0.08	1 9.11
June	8 7.32	6 9 11	1 10.21
July	8 5.71	6 4 05	2 1.66
August	7 10.83	6 0.11	1 10.72
September	7 7.02	5 10.78	1 8.24
October	7 10.91	6 0.64	1 10.27
November	8 2.99	6 3 39	1 11.60
December	9 5.10	7 2.21	2 2.89
Annual mean	9 3.18	7 5.72	1 9.46

In regard to the objection which might be made that the discharge of a stream is not in proportion to the gauge-readings, because in course of time changes take place in its bed, Dr. Berghaus remarks that in the lower Elbe, except in a few stretches, a raising of the bed has generally taken place. This being the case, if the discharge remained the same, a raising of the water surface or increase in the annual mean of gauge-readings should result, whereas, according to the preceding table, there has been a notable decrease.

Dr. Berghaus also publishes in his hydrography the tables of gauge-readings taken at Küstrin on the Oder between 1778 and 1835, from which it results that if we divide the whole period of observations of 58 years into two half periods of 29 years and calculate the mean stages, that the mean of floods has decreased 9.4 inches; of lowest stages 9 ft. 9.85 inches, and the annual mean 10.55 inches in the last half period, as compared with the first.

After giving the results obtained from the tables of gauge-readings taken on the Elbe and Oder the most mature deliberation, Dr. Berghaus was convinced that the discharge of both of those streams had diminished considerably, and he expressed his apprehension that both of those German rivers would disappear from the list of navigable streams if their discharge continued to decrease at the same rate as has been established for the period since 1781.

In order to obtain a clear conception of the relative water stages of the Elbe at Magdeburg between 1782 and a very recent period, and also to ascertain if the apprehensions expressed by Dr. Berghaus were being fulfilled, I have added to the tables furnished by him those of gauge-readings taken by Mr. Maass, inspector of hydraulic works, from 1836 to 1869, and have plotted on Sheet 3 the highest and lowest readings, as well as the mean of each year. I also divided the whole period of 142 years into three, the first two each of 50 years, and the third of 42 years, calculated the arithmetical mean of the readings for each of them, and represented them on the same plot.

The following very interesting phenomena in the relations of discharge of the Elbe at Magdeburg during this period of observation of 142 years are apparent from this plot and the calculated mean of water stages.

1. During the first period of time, from 1728 to 1777, the height of floods differed but little in the different years, and the highest only twice went up +17' 7.65" and +17' 9.2".

During the second period, from 1778 to 1827, the difference in the height of floods in different years already became greater. They went up ten times to between +17' 7.65" and +18' 5.76", and yet the annual mean of these floods was 12.45 inches less than that of the preceding period.

During the third period, from 1828 to 1869, the difference in the height of floods in different years became still greater; the latter, although the period was eight years shorter than the preceding, went up to from +17' 7¾" to +18' 8" at ten different times, and the arithmetical mean of all the flood heights was 3¼ inches greater than that of the preceding period.

2. It appears from this that in the last two periods floods occurred in the Elbe oftener and rose higher than during the first period. From this the conclusion can be drawn that in the latter two periods there was a greater discharge of the Elbe at high stages during some years. This clearly results from the circumstances that the waters which fall during heavy rain-storms flow off more rapidly from the barren mountain sides, and fill up the bed of the stream oftener and to a higher point than was the case during the first period.

3. The lowest water stages of the separate years went down, and as follows, viz:

	Feet.	Inches.
In the first period to	+3	8
In the second period to	+2	0
In the third period to	+1	5.65

and the mean of the lowest stages was 19.71 inches lower during the second than the first period, and 10.37 inches lower during the third than the second, and consequently 30.08 inches lower during the third than the first.

From this will be seen that the lowest stages of the Elbe now occur much oftener and sink considerably deeper than in the last century, and this establishes the fact that here a considerable sinking of the water surface of this stream has taken place.

4. The greater differences between the lowest and highest stages amounted as follows, viz:

	Feet.	Inches.
During the first period	14	1.65
During the second period	16	6.95
During the third period	17	9.73

These increasing differences in the stages are the result of a deeper sinking of the lower stages and a higher rising of the floods.

5. The arithmetical means of the annual gauge-reading was 17.76 inches lower for the second period than the first; 14.63 inches lower for the third than the second; and consequently 32.39 inches lower for the third than the first period.

It is a fact that at higher stages the cross-section of discharge, as well as the mean velocity of the current, increases considerably, while at low stages both decrease in a notable degree, and that therefore a foot in the rise of high stages represents quite a different quantity of discharge than a foot in fall of the low stages.

Nevertheless the means of the annual gauge-readings can be taken as approximately representing the relative discharge of those years, because the flood stages generally only last a few days, while the lower stages last much longer, and because the medium stages last during the greatest part of the years and have the greatest influence on the annual mean.

The correctness of this opinion is confirmed by the plot on Sheet 2, of the annual stages and discharge of the Rhine at the Germersheim cut-off, since the annual discharge calculated by Mr. Grebenau, who gave due consideration to the increase in cross-section of discharge and mean velocity of the current during high and the contrary during low stages, agrees very nearly with the discharge due to the annual mean.

We can therefore draw the conclusion from the decrease in the annual means, as shown .above under the head of 5, that the discharge of the Elbe at Magdeburg was less during the second than the first, and again less during the third than the second period, and that therefore a continual decrease in the discharge of the Elbe during the long period of 142 years is proven.[*]

Above all that has just been said, it should be stated that such thorough observations and studies have been made of the condition, the relative discharge, and the navigability of the Elbe since 1842 as have probably never been made on any other river in Europe. It was agreed by the Elbe navigation acts, which were adopted in 1842 by the eight states bordering on this stream, whose representatives were assembled at Dresden, that the hydraulic engineers of these eight states should frequently navigate and examine the river; survey all obstructions to navigation; observe the effects of all completed works of improvement, and suggest the works which were still necessary to remove existing obstructions.

This commission on the inspection of the Elbe, composed of the most distinguished hydraulic engineers of the states bordering on it, while taking observations on it with a view to determine the conditions, the navigability, and the improvement of its water route, so important to Germany, collected very interesting data on the changes in its bed and its relative discharge, which are described in detail in their published proceedings submitted to their central governments.

I submit herewith only a brief extract from these proceedings of those facts which were deduced and established which have a bearing upon the relative discharges of the stream.

1. The first technical commission on the inspection of the Elbe, which navigated and examined the stream between September 6 and October 15, 1842, established the fact—first, that the low-water stage during August and September, 1842, was the lowest that had ever occurred.

---

[*] According to my calculation and plot, the difference of the annual means of the first and second period is 3.76 inches less than according to Berghaus' calculation. This arises from the fact that I included the years 1728–1730 in the first period, and thus made a different subdivision, and that I eliminated the low gauge-readings which were solely due to ice gorges, and substituted others furnished by Mr. Manss.

That Mr. Manss found the decrease in the lowest and annual mean stages greater by 4.7 inches to 5.22 inches than I did is explained by the circumstance that he subdivided the period of observation for the purpose of comparison into three periods of 23, 67, and 53 years respectively, so that his figures apply to the period of 120 years, the total of the last two, while mine apply to one of 92 years only.

This was apparent not only from the tables of the readings of the various gauges on the Elbe, but particularly from marks and figures which had been cut into the rocks in the river to indicate the lowest stages which had occurred.

These marks were first found on the three large rocks at Tetschen, with the dates 1616, 1719, 1766, 1782, 1790, 1800, and 1835. The low-water marks of these years were all from 4.69 inches to 8.85 inches higher than that of 1842. Low waters are marked on the large flat rock in the bed of the stream at Pirna for the years 1616, 1706, 1707, 1746, 1834, and 1835, and all of these are from 5.21 inches to 11.46 inches higher than the one of 1842. On the rocks lying in the bed at Strehla, also, the lowest water stages are marked for 1718, 1746, 1790, 1800, 1834, and 1835, and these are higher than that of 1842.

2. The members of this commission then agreed upon the following, viz:

It having been ascertained, from the table of gauge-readings taken between 1811 and 1842, that the lowest stage of the years 1811 and 1835, which were distinguished on account of their extraordinary drought, are nevertheless 6.25 inches higher than that of 1842, and that all the marks which were found to indicate previous extraordinary low stages, are universally higher, that a line of reference (that is, comparative water-surface line) from which the depths of the water to be obtained shall be compared, shall be adopted which shall indicate a stage 6.25 inches higher than the lowest of 1842.

This commission also recommended to the states that the depth in the upper and middle Elbe should be so improved by suitable works that at the assumed lowest stage, 6.25 inches higher than the lowest stage of 1842, it should be sufficient for navigation by half-laden vessels of the size at that time in use. These drew about 32.93 inches. The required depth was therefore about 37 inches.

Both of these recommendations were adopted by the states, and arrangements were made to build works of improvement on the Elbe throughout its whole length which would at ordinary low stages give the least normal depth of 37 inches.

3. At the subsequent examinations of the stream in 1858 and 1869,[*] the hydraulic engineers were surprised to find that the longitudinal profile of the Elbe had changed considerably in the relatively short period of time since 1842. The lowest stages which occurred during this period were in August and November, 1857, and September, 1869. In the upper Elbe, that is, in the kingdoms of Bohemia and Saxony to below Dresden, these were 6.98 inches lower than that of 1842, the lowest previously known, but in the middle Elbe they were 7.31 inches, and in the lower river 16.93 inches higher. This can be seen from the plot of gauge-readings taken on the Elbe and shown on Sheet 4.

The Austrian and Saxon commissioners proved in detail and conclusively, in the opinion expressed by them in the proceedings of 1858, that the sinking of the low water of 1857 below that of 1842 could not be ascribed to a general deepening of the bed in that part of the stream, because the latter was confined at several places between rocky contours; was established by the numerous mill-dams; and finally since its cross-section of discharge is made unchangeable at the old bridge of arches at Dresden by the pavement of dimension stone with which the sole of its bed is covered.

The lowest stage of 1857 at this bridge was 8.36 inches, and all the gauges up to Melnik were on average 6.98 inches lower than that of 1842. This can only be explained by the circumstance that the discharge of

* The proceedings of the inspecting commission in 1850 could not be examined by me, since they were unfortunately lost.

the Elbe must have been considerably less in August and September, 1857, than it was during the lowest stage of 1842, in consequence of the great drought which prevailed in that year.

During the low stages of 1857, while the discharge of the Elbe was less, the simultaneous readings at the fixed gauges on the middle river between Meissen and Torgau averaged an increase of 7.31 inches, and those on the lower river as far as the head of tide-water 16.93 inches above those of 1842. In the well-based opinion of the above-mentioned members of the commission this can only be explained by the circumstance that in these portions of the river the water-surface was considerably raised by local deposits of heavy material which act like dams.

The same conclusions apply to the phenomena which were observed in 1869 as in 1857; the low-water stage in the upper Elbe was as low as that of 1857, and yet in the other portions of the stream it was higher than that of 1842. There was a smaller discharge, but the water-surface was raised in the middle and lower Elbe by the filling up of the river bed, and consequently the gauges there gave a higher reading than in 1842.

The six members of the commission of 1869, basing themselves upon those facts, agreed unanimously that in their opinion the longitudinal profile of the Elbe had changed very much since 1842, and that the bed of the river in its lower portions had raised considerably, and that therefore the *depths* to be obtained should no longer be determined by the low water of 1842, but, in consideration of the changed character of the longitudinal profile, by the lowest stages observed in the separate portions of the stream.

4. These very important data collected by the hydraulic engineers of these Elbe states establish the fact that in September, 1842, the river was at a lower stage than it had probably ever been during centuries, and that it already went lower in 1852, 1857, and 1869 in its upper portion, and it is perfectly clear therefore that since 1842 there has been a considerable decrease in the low-water discharge of the Elbe.

5. The correctness of these conclusions is also strengthened by the fact that the normal widths of the improved river, which were determined by an international commission upon data collected even before 1842, and which were adopted and carried into execution, and which were declared to be suitable for the purpose by the inspecting commission of 1842, have been gradually narrowed. Up to 1869 this narrowing amounted—

In Bohemia	From	494½	to 321½	feet.*
In Saxony	From	417½–556	to 370½	feet.
Below the Saxon frontier	From	432½	to 278	feet.
In the Duchy of Anhalt	From	556	to 494½	feet.
Below the mouth of the Saale	From	618	to 556	feet.
Below the mouth of the Havel	From	803	to 741½	feet.
From Schnakenburg to Dörnitz	From	865	to 803	feet.
From Hitzacker to Radegast	From	1,098⅔	to 889¼	feet.
&c., &c.				

After this inspection of 1869 the commission also expressed the opinion that in carrying out further works of improvement it seemed advisable to decrease the normal widths still more, since they found in many improved portions of the stream large sand-bars, which caused troublesome bends in the channel and seemed to indicate that the width taken was too great.

---

*A "Rheinländische ruthe" is equal to two "klafter." A "klafter" (or fathom) is taken at 6.178 English feet.

Since the normal widths for the improved Elbe which were adopted at the beginning of the work, and which suited the conditions of its discharge during a long period of years, have lately been shown to be too large, and their decrease has become imperatively necessary, the conclusion is fully justified that the discharge of this river at medium stages is less than it was before 1842, and that there has been a general decrease or lessening in its water consumption.

6. The hydraulic engineers comprising the first commission of 1842 held the following opinion concerning the capacity of the river, viz:

That, although we have no material on hand to enable us to treat this subject in an exhaustive manner by the comparison of figures, yet it is proper to state that a marked recent decrease in the discharge of the stream is not indicated by experience or observation. But we recognize the fact that the clearing of forests, cultivation of marshes, and irrigation of lands are able to produce a marked decrease in the quantity of water in the stream.

As it is generally known that between 1842 and 1869 many forests have been cleared, extensive marshes cultivated, and irrigation of meadows carried out on an extensive scale in the valley of the Elbe, it follows naturally that a marked decrease in its discharge has taken place, just as the commission foretold.

7. The commission of 1842 also declared, with reference to the deposits brought down by floods:

No experiences nor observations are known to us which would substantiate the assumption of a progressive and visible general raising of the bed of the Elbe. Moreover, it is true that from such of the observations which have been made, and which seem reliable and applicable to the question, it may be assumed that no such raising has taken place.

The commission of 1858 already proved a partial, and that of 1869 a general filling up and raising of the bed of the middle and lower Elbe, which evidently must have taken place during the later period of 27 years, since very large quantities of earth, gravel, and sand were washed down from the cleared mountain sides, and since considerable masses of gravel and sand were forced down by the improvement and narrowing of the river, which were executed with great energy and on an extensive scale in Bohemia and Saxony. These deposits remained in the middle and lower stream and caused the raising of the river bed, because the propelling power of the water is there less in consequence of its slighter slope, because the improvements had not been undertaken on so extensive a scale as above, and because the method of improvement by means of spur-dikes adopted here does not seem so effective nor suitable as the method of training-walls adopted above.

8. It will be seen from the tables contained in the record of proceedings of the various commissions in the inspection of the stream, that independent of the numerous works for the regulation and correction of the Elbe which existed prior to 1842, that the states bordering on it had expended in the 27 years, including 1869, for works for the regulation, correction, and narrowing of the stream: for the preservation of these works, and finally for clearing and deepening of the bed of the stream from Melnik to the sea, a distance of about 639 miles,* the noteworthy sum of twelve million dollars.† With this sum about 86¾ miles of paving, 68 miles of training-walls, and 5,241 spur-dikes were constructed and maintained. Every one will therefore admit that the Elbe states have done a very great deal for its regulation and for the improvement of its navigation.

If now we read the records of the proceedings of the commission of 1869, in which appears a detailed description of the condition of the

* A "Deutsche meile" (German mile) is here taken at 10,126 English yards.
† A German "gulden" is taken at 40 American cents.

stream and the obstructions to navigation which still existed, it will be found that at 124 points there was only the insufficient depth of from 18¾ to 31¾ inches in the channel, and that at 113 points navigation encountered great difficulties, partly from insufficient depth of water and partly from the sharp bends. At nine of these points real interruptions of navigation take place. Vessels get aground and change the channel so that both ascending and descending vessels are often detained a long time, and as a rule are eventually pulled over these dangerous places with great trouble and at great expense in money and loss of time. It will further be found from the records that the light-draught passenger boat, on which the commission navigated the river frequently, was stopped by the jam at these places and assisted in pulling the other vessels over in order to get past, and frequently ran aground. No account of such glaring obstructions to navigation are recorded in 1842 nor in 1858.

It therefore appears that the work which has been carried on during 27 years at so great an expense has not improved the channel of the Elbe in general, or at least to any considerable extent, and has even made it worse in several places.

These results might lead many engineers to the conclusion that either hydraulic art in general, or at least the system of works adopted on the Elbe, is not suitable to improve a channel of a stream even with the expenditure of considerable money. This conclusion would be very erroneous, since the unsatisfactory results obtained by the works which have been carried out are easily accounted for. They were produced by the decrease in discharge which has taken place since 1842, and the immense bars of gravel and sand which were carried and deposited in the middle and lower Elbe, and which caused the channel to get shoaler, narrower, and more crooked. An additional trouble was that the works which were built in former years, and at great expense for the purpose of confining the width of the stream to the established normal width, gradually ceased to be of any effect since this normal width was found to be too great in consequence of the decrease in the discharge.

9. The proofs of the correctness of the conclusion which I have heretofore drawn from the gauge-readings taken at Magdeburg, that the discharge of the Elbe is gradually decreasing, are unassailably strengthened by these experiences gained and data collected by the commission on the inspection of the Elbe, and in addition thereto, some of the phenomena which appear on the plot of the gauge-readings are made clear.

From the plotted gauge-readings of the Elbe taken at Magdeburg, we have seen that the floods in later years have risen considerably higher than formerly; that however the arithmetical mean of the heights of all the floods during the second period from 1778 to 1827 were 12½ inches less than that for the first period from 1728 to 1777, and that, on the contrary, this mean for the last period from 1828 to 1869 was again 3½ inches higher than that of the second period.

Now it appears from data collected by the commission that since 1842 the normal width of the stream for medium stages was reduced 61¾ feet above and below Magdeburg, and that the lowest readings taken at that station are from 13½ inches to 20¾ inches higher than they were in 1842, due to the rising in the river bed, produced by the deposits which were made in it. It must therefore be admitted that these marked changes in the river bed since 1842 must also have contributed to increasing the number and height of the flood stages.

These data furnished by the commission also contradict the statement of Mr. Maass that the bed of the Elbe has been deepened considerably during later decades, and disapproves his conjecture that this deepening has caused the decrease in the height of the gauge-readings.

If the filling up of the river bed, which the commissions have shown to have taken place since 1842, had not occurred, the lowest stages at Magdeburg would have gone down at least 16⅔ inches, and the annual means since 1842 would have become far less, and the decrease in the height of the lower stages, and annual mean of the period from 1828 to 1869, which is shown by the plotted gauge-readings to have been from 10½ inches to 14½ inches, would certainly have been much greater than that which took place in the period from 1778 to 1827. This proves that the discharge of the Elbe since 1728 has decreased steadily, and during the last three decades even more rapidly than formerly.

I have treated the hydrographic conditions of the Elbe at such length in this paragraph because I had at my disposal exact gauge-readings taken on it during the longest period of time, and at the same time the valuable data collected and experiences gained by the inspecting commission. From them the most valuable conclusions could be drawn relative to the conditions of discharge of this river, and which may be considered applicable to other analogous streams.

THE VISTULA.

Mr. Schmid, of Marienwerder, royal Prussian privy counsellor of the government, gives, in the Journal of Architecture of Erbkam, 1858, detailed information concerning the hydrographic conditions of the Vistula, and publishes in the treatise also the gauge-readings taken at Kurzebrak and Marienwerder during the period of 48 years from 1809 to 1856, inclusive, together with very interesting tables showing the conditions of discharge of this stream, and gives the following as his opinion, viz:

That the apprehensions which were created in the country by the extraordinary flood of 1855, that larger quantities of water flow into the Vistula during recent than in future years, are not grounded, and that the tables rather indicate that a decrease in the quantities and stages of the water may have taken place.

I have supplemented these tables with the gauge-readings taken up to the end of 1871, and which were furnished me in a most friendly manner by the Royal Prussian government officials, and have plotted these gauge-readings taken during a period of 63 years, (on Sheet 5.) I have divided the latter into the two, from 1809 to 1840 and from 1841 to 1871. From this plot the following is apparent, viz:

1. During the first period the floods only rose higher than +20 feet 7 inches in 5 different years, and the highest reached +24 feet 0.18 inches, whereas during the second there were 9 years in which they went up above 20 feet 7 inches, and the highest reached +28 feet 5.62 inches. From this it appears that the apprehensions of the inhabitants that at present during floods a much larger quantity of water flows into the Vistula, and that these now produce more numerous, higher, and consequently more destructive inundations than in the former period from 1809 to 1840, are proven fully grounded and true.

2. During the first period there was only one year in which the lowest stages went down to +9.27 inches, while during the second they went down to between −4.12 inches and −1 foot 11.68 inches, so that the arithmetical mean of the stages during the second period was 2 feet 4.6 inches less than that of the first. From this reason it may be asserted that a considerable depression in the low-water stages of the Vistula has taken place.

3. The difference between the highest and lowest stages was 21 feet 3.05 inches in the first period and 30 feet 5.54 inches in the second, an increase of 7 feet 2.49 inches.

4. The arithmetical mean of all of the gauge-readings taken during the second period is 1 foot 5 inches less than that of the first period; from which it can be concluded, by comparison with similar observations made on the Elbe, that the discharge of the Vistula at low and medium stages has decreased considerably during the period of observations of 63 years.

It will be seen from the above that all of the phenomena observed on the Elbe are found on the Vistula, and even in a more marked degree.

THE DANUBE AT VIENNA.

In order to obtain a clear presentation of the conditions of discharge of the Danube at Vienna, and to utilize as much as possible the knowledge of these in the improvement of the river, with which I had been intrusted, I caused the gauge-readings for the highest and lowest stages, and the calculated means of monthly and annual readings, taken at the great Danube bridge to be tabulated.

Since these gauge-readings taken in this important portion of the river alone are interesting and acquire a greater interest in the comparison of the changes which it is evident will be caused by the improvement, I considered it my duty to publish them in tabular form at the end of this paper. I have at the same time, for the sake of perspicuity, plotted them on Sheet 6.

I must here remark that the recorded gauge-readings only begin at 1826, those of former years having been unfortunately lost. I could therefore only present them in my tables for the period of 46 years, from 1826 to 1871.

Up to the present time ice-gorges have frequently been formed in the unimproved portion of the Danube at Vienna during the winter months. These so change or fill up its bed that the water above them is raised considerably even at low stages and during dry periods and produce inundations at Vienna. Since these backwaters are not really flood stages, I have when they occurred (denoted on the plot with dotted lines) substituted the reading for the stages which would have occurred had the gorge not been formed. I obtained these true readings by interpolating the readings of the gauges next above and below, and have entered them in the tables as well as on the plot.

If now the period of 46 years is divided into two equal periods in these tables and in this plot and the mean heights are calculated for them, the following phenomena will appear, viz:

1. During the second period floods rise higher and occur oftener, and, on the contrary, the lowest stages in the river occur more frequently and go down lower than during the first.

2. During the second period the calculated mean of flood stages were 10.37 inches, that of all stages 8.71 inches, and that of the lowest stages 5.19 inches, lower than during the first.

3. The decrease in the mean annual stages was very different for the separate months of this year, and it varied during the five months, from April to August, from 1.03 inches to 8.3 inches, and during the seven months, from September to March, from 9.34 inches to 1 foot 9.78 inches.

I must further mention the following facts connected with the decrease of gauge-readings at Vienna.

If we divide the period of observation of 46 years into five and add together separately the number of days in each period at which the readings were above and below zero, the following will be the result, viz:

Average number of days during each year.

Periods.	Below zero.	Above zero.
First period	31	334
Second period	108	257
Third period	102	263
Fourth period	137	228
Fifth period	162	203

From these relative figures it will be seen that the number of days in which the stages are above zero steadily decrease, and, on the contrary, those in which they are below rapidly increase.

I could only find the following starting points to ascertain in the conditions of discharge of the Danube during the last century.

I found among the archives of public works in the Imperial and Royal Northeast City Hall an accurate cross-section of the Vienna Danube Canal, taken at its head at Nussdorf during the last century, but unfortunately without date, on which were indicated the highest, mean, and lowest stages which occur there, and the following appears in quotation marks, viz:

The highest stage rose to 8 feet 3.58 inches above the mean, and the lowest fell 4 feet 1.79 inches below. The greatest variation therefore amounts to 12 feet 5.37 inches.

At present the highest floods at the head of the Vienna Danube Canal at Nussdorf rise to $+16$ feet 7.16 inches, and the lowest fall to -4 feet 7.92 inches. The greatest difference is therefore 21 feet 3.08 inches.

In my lecture, which I delivered on March 11, 1871, before the Society of Austrian Engineers and Architects, on the improvement of the Danube at Vienna, I pointed out the fact that the floods at Nussdorf rose 4 feet 1.79 inches higher than formerly. This is caused by the narrowing of the river bed at that place by the construction of the spur-dikes on the left bank. It follows then that the low-water stages at that point are 4 feet 8 inches lower than during the last century.

In order to convince myself that this sinking in the stages, particularly of low water in this vicinity, did not result from a deepening of the bed, I compared the reading of several gauges in Lower Austria with those taken at Vienna, and tabulated them in the form shown at the end of this treatise.

It will be seen from this table that during the period of observation anterior to 1854, when the water was at the zero at the great Danube bridge at Vienna, the other gauges showed a quite different reading, some of them being as much as $+2$ ft. 1 in., and others -1 ft. 8.75 in.

By virtue of a decree from the Imperial and Royal Ministry of Commerce, No. 7062, issued September 9, 1853, the zero of all the other gauges on the Danube in Austria were set exactly to the same point as the zero at Vienna. This was done during a long-continued zero stage of the river at the latter place in the fall of 1854, and by moving the graduated scales up or down.

The readings taken between 1855 and 1871 show, however, that at every zero stage which occurred at Vienna nearly all the reading of the other gauges, both above and below, were higher and increased from year to year. These readings varied between $+4.15$ inches and $+4$ feet 4.9 inches. I have shown these changes in the longitudinal level of the stream by the plot of the present zero stage in Fig. 3, Sheet 6. This striking change in the zero stage at the various gauges can only have been caused by the improvements which have been made at Vienna since 1850, and which, by narrowing the stream, have caused a deepening of its bed and sinking of its water-surface. Similar deepening of the bed and sinking of the water-surface are also observed at the gauges at Linz, Wallsee, and Tulln. At all the other portions of the river the bed was raised by deposits and the water-surface raised thereby.

Since the bed of the stream at Mölk and Stein consists partly of rock, and at the latter place the river is confined to its normal width by the bridge which exists there, and since in these two portions of the stream no deposits are observable, the bed of the river may be considered to have been unchanged there since 1854.

The readings of the gauges at these two points have been alike during recent years, when that at the great Danube bridge was at zero. They were +1 foot 3.56 inches. The readings at the Ferdinand's bridge across the Vienna Danube Canal, at which the gauge had also been set to the same zero as that at the great Danube bridge in 1854, are now from 1 foot 1.48 inch to 1 foot 5.63 inches higher than at the latter. It may be concluded from this that (in consequence of the deepening of the bed) the low-water stages at Vienna have been lowered 1 foot 3.56 inches. At the next gauge above, at Nussdorf, this depression yet amounts to 11.41 inches. On the contrary, in consequence of deposits, the low-water surface in other portions of the river, as at Grein, Strudeu, Fischamend, and Hainburg, have been raised from 1.45 inch to 3 feet 1.35 inch.

Although this deepening of the river bed at Vienna is made perfectly clear by the plot in Figure 1, which shows a decrease in the low and medium gauge-readings of from 5.19 inches to 8.71 inches during the period from 1849 to 1871, I was also fully convinced that the discharge of the Danube at Vienna had also decreased during the last decades. This conviction resulted from the fact that during the first period of observation, from 1826 to 1848, the Emperor's stream, about 497 feet 9 inches wide, still discharged quite a considerable volume of water even at the zero stage, while during the last decades the discharge at that stage of this subsidiary stream, which has been much filled with sand, amounts to but very little.

The gauge-readings taken at Old Orsova furnished me with the most striking proof of the observations which I had made that the discharge not only of the Danube but of the most of its larger tributaries had decreased considerably during recent periods, and I therefore take the liberty of commending the data gathered from them, and which I will now enumerate, to the special consideration of my readers.

I delivered a lecture on the improvement of the Danube at the eight rocky reefs and rapids near Orsova, and illustrated it with plans and longitudinal sections. These were published in the Journal of the Society of Austrian Engineers and Architects, (No. 9, 1872.) From the latter it will be seen that the bed of the Danube in the whole of the mountain gorge, a distance of about 75½ miles* from Bazias, to about 5.89 miles below Old Orsova, consists almost entirely of rock. This portion of the river bed has remained unchanged during centuries. †

The first Imperial and Royal Private Danube Steam Navigation Company erected a gauge in 1838 at Old Orsova, one of its most important

*An Austrian mile = 4,000 Austrian klafter = 8,295⅘ English yards.

†The Danube Steam Navigation Company from 1847–49 and the Austrian government in 1855 did, it is true, remove some dangerous projecting rocky reefs in the lower part of the Iron Gate. The relations of discharge were, however, not altered thereby at the cross-section at Orsova, 5.89 miles above, because the upper rocky reef in the Iron Gate, which extends across the whole bed of the river, remained unchanged.

landings, where readings were accurately observed and recorded, since their freight rates were partially regulated by them.

The gauge-readings at Old Orsova are especially suitable to determine the conditions of discharge from the whole basin of the upper Danube for the following reasons, viz:

All of the large tributaries have already emptied into it; it has a bed at Orsova which is regular and sufficiently wide for a long distance; it is there already such a mighty stream that the conditions of discharge are not visibly affected here by abnormal stages arising from abnormal discharges from the basins of the different tributaries which are caused by actions of the elements, and because the high and low stages of the larger tributaries occur at different periods of time. This happens in consequence of the great extent, different configuration, as well as the composition of the soil, and difference in climatic conditions of their basins.

The Danube at this point may therefore already be regarded as a great regulator of the abnormities which occur in its different tributaries, and the gauge readings here may therefore be taken with perfect confidence as quite nearly expressing the relative conditions of discharge from the whole basin of the Danube.

After eliminating the incomplete records of the year 1838–1839, I compared the accurate readings at the Old Orsova gauge during the 32 years from 1840-1871, in the table at the end of this paper, and from this made two plots on Sheet 7,—the one showing the highest and lowest stages of each year and the calculated annual mean, the other the calculated arithmetical mean of the monthly and annual stages. From these the following is visible, viz:

If the period of observation of 32 years is divided into two of 16 years and the mean stages calculated for them, we will find, viz:

1. That the flood stages of the Danube at Orsova occur, it is true, oftener during the second than the first period, but do not rise so high, and that the arithmetical mean of these stages during the second period is 11.41 inches less than that of the first period. This proves that now, even at flood stages, the discharge of the Danube at Orsova is less than it was during the first period of observation.

This phenomenon is quite different from those which we have already observed on the Rhine, the Elbe, and the Vistula, and is explained by the following: The heavy rain-falls, and the melting of the snows in the basins of the larger tributaries, and the consequent floods in the latter, throughout the extensive basin of the Danube, occur at different periods of time, and therefore very often simultaneously with the low and medium stages of other tributaries. From this it follows that the flood stages at Orsova may be considered as the already adjusted mean of flood stages, and this proves that the increase in the discharge occasioned by floods in some tributaries cannot compensate for the decrease occasioned by the mean and low stages in the other.

2. Low stages occur oftener in the second period, and sink in some years 2 feet 4 inches lower, so that the mean is 1 foot 3.21 inches lower than during the first. From this it may be concluded that a considerable sinking in the water-surface of the Danube has taken place at Orsova, and that there at present the discharge is very much less than during the first period of observation.

3. The annual mean decreased 1 foot 6.15 inches during the second period, and therefore the total annual discharge of the Danube at Orsova has decreased during that time.

4. This decrease in the annual mean is quite different during different

months, and amounts to only from 2.04 inches to 1 foot 1.48 inches from January to April, but on the contrary to from 1 foot 1.59 inches to 2 feet 3.18 inches from May to December.

5. The decrease in the gauge-readings of the Danube at Orsova has been relatively much greater during the short period of observation of 32 years than that observed at the streams heretofore mentioned, and for the reason that several large tributaries, such as the March, Waag, Drau, Save, and Theiss, have an effect on the lowering of the water stages and decrease in discharge.

6. These plots of the gauge-readings show that the highest, lowest, and calculated monthly and annual means of the Danube at Orsova have decreased considerably,—this decrease being synonymous with a general sinking of the water-surface of the stream, and this could only have taken place in this rocky bed, which has remained unchanged during centuries, in consequence of a considerable decrease in the discharge of the river. This strengthens the conclusion, which I have heretofore drawn, that the decrease in the gauge-readings of the lower and medium stages on the Rhine, Elbe, Oder, Vistula, and Danube at Vienna, which I have heretofore shown has taken place, even after deducting several inches due to the effect of recent improvements on these rivers, could only have resulted from a continuous decrease in their discharge.

After the foregoing had strengthened me in my belief that the discharge of the Danube at Old Orsova and also at Vienna had decreased and would still further decrease, I was impelled to submit marked changes in my two projects for making the Danube navigable at the seven rocky reefs of the Iron Gate above Orsova and the improvement of the river at Vienna.

In the project which I submitted in 1854, to blast channels through the eight rocky reefs in the Iron Gate and above Orsova, I fixed t bottom of these channels at 6 feet 2.76 inches, or 7 feet 3.11 inches below the water-surface of the zero stage. The Imperial Royal Danube Steam Navigation Company and its engineers declared that a depth of 4 feet 1.38 inches below that stage would be sufficient. Based upon the results of the investigations which I have enumerated, I have now submitted to the government the necessity of blasting these channels to at least 8 feet 3.56 inches, or else in about 30 years they will no longer be navigable at low stages.

In the first project for the improvement of the Danube at Vienna which was approved, the experts in the year 1867–'68 fixed the normal width, measured between the upper edges of the sides of the cut-off for the cross-section of discharges at low and medium stages at 1,244.5 feet.

Referring to the decrease in the heights of stages and in discharge, I then proved by means of hydraulic calculations that the above width was clearly too great, and that a width of 908.5 feet would be more suitable for the discharge and future slope.

The seven hydraulic experts which were invited by the commission on the improvement of the Danube to give an opinion on this matter did not consider that the decrease in the discharge of the river at Vienna mentioned by me as yet a clearly-proven fact, but in view of all the circumstances mentioned by me and in consideration of the evils which are now showing themselves on the improved Rhine between Basle and Mannheim, and which are clearly due to the fact that its normal width was taken too large, they reduced the normal width for the improved Danube at Vienna from 1,244.5 feet to 1,120 feet. The Danube commission accepted this, and ordered the execution of the improvement with that normal width.

I consider this reduction as the first important result of my proofs of
the decrease in the discharge of streams, since I am convinced that if
the width of the new bed of the stream had been retained at 1,244.5 feet
that, just as was the case on the improved parts of the Rhine and Elbe,
there would have been formed in it numerous sand and gravel bars and
some of these even on the right-hand wharf sides; that the channel between these bars would have become very crooked; that the crossings
from one concave bend to the other would have become very shallow,
and that the whole expensive improvement would have been declared
faulty and abortive.

In order to enable me to compare more readily the decrease in the
height of stages in the five above-mentioned rivers, I reduced it to the
same period of 50 years, and obtained the figures which are contained
in the following table:

Name of the stream and gauge-station.	Periods of observation and their duration in years.	Decrease in the height of medium stages during the half period of observation.		Increase or decrease in the mean flood stages during the half period of observation.	Decrease in the height of the mean stages reduced to a period of 50 years.	
		Of lowest stages.	Of annual stages.		Of lowest stages.	Of annual means.
		Inches.	*Inches.*	*Inches.*	*Inches.*	*Inches.*
Rhine at Emmerich	1770–1835, 66 years.	13.64	16.01	+ 0.85	20.65	25.62
Rhine at Düsseldorf	1800–1870, 71 years.	− 0.62	4.87	+ 8.84	− 0.85	6.86
Rhine at Cologne	1782–1835, 54 years.	7.42	4.40	+ 15.44	13.73	8.14
Rhine at Germersheim	1840–1867, 28 years.	unknown during 92 years.	17.12	unknown.	unknown.	61.16
Elbe at Magdeburg	1728–1869, 142 years.	20.86	31.92	− 9.27	16.23	17.35
Oder at Küstrin	1778–1835, 58 years.	9.73	10.43	+ 1.61	16.75	17.97
Vistula at Marienwerder	1808–1871, 63 years.	28.48	16.99	− 1.63	45.21	20.98
Danube at Vienna	1826–1871, 46 years.	5.19	8.71	− 10.37	11.73	18.94
Danube at Orsova	1840–1871, 32 years.	15.20	18.14	− 11.41	47.49	56.70

By far the greatest part of the sinking of the water-surface of the
Rhine at Germersheim is due to the deepning of its bed caused by its
extensive improvement by cut-offs, and only the smaller part is due to a
decrease in its discharge. If the gauge-readings at that station are
therefore eliminated from the above table it will be found that the
greatest decrease in height of water-surfaces has been between 47.49
inches and 56.70 inches on the Danube at Orsova, at Marienwerder on
the Vistula between 26.98 inches and 4.52 inches, at Emmerich on the
Rhine between 20.65 inches and 25.62 inches, and at the other stations
between 6.86 inches and 18.94 inches.

The decrease which is shown on the Elbe in the lowest and medium
stages as amounting to 16.23 inches and 17.35 inches would certainly
have been greater if sand deposits had not been made in its bed since
1842, which again raised the water-surface from 13.03 inches to 22.06
inches.

The sinking in the water-surfaces increases with the size of the stream
and the vicinity of the station to its mouth, since there the basins are
larger in extent and the sum of all the decreases in the discharge of the
tributary rivers and creeks above comes into play.

The plots of the gauge-readings show in regard to the floods in these
rivers (with the single exception of those taken at Orsova) that in recent

periods these occur oftener and rise higher than in former times. It follows clearly then that during flood stages the discharge is much greater than it was formerly.

These plots show further that the floods were more regular during the successive years, whereas during the last decades very high ones occur in one year and then in one of the succeeding years they do not rise so high, and this proves that now the changes between a very wet and very dry year occur oftener and are much greater than formerly. This is particularly noticeable in the table on the Elbe and Vistula.

The causes of these remarkable phenomena are plainly the following, viz:

In consequence of the clearing of forest, particularly in the mountain regions, the rain comes down in torrents, and there are more bursting of clouds than formerly; the rains are less soaked in by the cleared lands, run off into the creeks, rivers, and streams more rapidly, and fill these with larger quantities of water, and finally, the rain descending rapidly down the cleared mountain sides tear up the soil and fill up the beds of the creeks, rivers, and streams with earth, sand, and gravel, and thereby increase the height of floods.

The correctness of this assertion is sadly substantiated by the inundations which are continually of more frequent occurrence in Italy, Southern France, Hungary, Bohemia, and many other countries.

These phenomena attending floods induced a distinguished meteorologist to express the conjecture to me that the increased discharge of present floods might possibly compensate for the diminished discharge at low and medium stages.

This conjecture is however entirely unfounded, since, as I have already shown on the Rhine and Elbe, the annual discharge is very closely represented by the discharge due to the annual mean of gauge-readings, and this has been shown to be decreasing for the five rivers which have been mentioned. The incorrectness of this conjecture is particularly shown by the table of gauge-readings at Orsova. It happens very often that, in the very extensive basin of the Danube, the floods on several of the larger tributaries occur at the same time as the low water of others, and yet there is no compensation for the diminished discharge of some by the increased discharge of others, since just at Orsova the means of the highest, medium, and lowest stages have decreased more rapidly than on the other four rivers.

Even if for some years the conjecture was really true that the continued extraordinary floods did compensate for the more frequent less discharge, this would be little consolation for mankind. For the disadvantages which arise to the consumption of water when the rise is at its low and medium stages would not only not be removed by these constantly increasing higher floods, but, on the contrary, increased by the more numerous and destructive overflows which now occur.

Having now furnished unassailable proofs that the lowest and medium stages, and therefore the discharge of the five principal rivers of Central Europe, whose total basins have an area of about 596,751 square miles, have been continually decreasing during a long series of years, the following conclusions may be drawn, viz:

1. Since these five streams are almost exclusively fed by the tributary rivers and creeks, the low and medium stages and discharge of the latter must have continually decreased during a long series of years.

One may easily be convinced of this by comparing the gauge-readings taken on them and tabulating and comparing them by the same method which I pursued with the five streams.

The correctness of the assertion is further substantiated by the fact that many manufactories established about 50 years since on the banks of creeks and rivers, which were then rich in water supply, are now sensibly feeling the decrease in the latter, which flows into their flumes, and are now erecting steam-engines as auxiliaries to their former abundant water-supply.

2. It is probable that the same causes which have acted in the basins of these five streams have also been at work in those of the most of the other rivers and streams of Europe, and, indeed, in the inhabited and cultivated portions of the three other parts of the globe, and consequently it may have happened that a similar decrease in the discharge at low and medium stages has taken place on most of the rivers on the earth, and that their floods now occur oftener and rise higher, and thus by their increased discharge cause more destructive overflows than they formerly did.

3. If the causes which have been at work during the last 140 years, and which have decreased the ordinary discharge of streams and rivers and again rapid filling up by floods, should continue to act in the future, it is clear that the low and medium stages of streams, rivers, and creeks will decrease in height still more, and the question will force itself upon the mind of everyone to what extent this decrease in their discharge may go.

There need probably be no apprehension that the low water-surface of the Danube, Rhine, Elbe, and Vistula will ever go down to their beds, that is, that they will become partially dry, because the first two are partially fed by the ice and snow of the Alps; because the causes which create this decrease will probably not act beyond a certain point, and because the many tributary creeks and rivers which empty into these streams generally have their highest and lowest stages at different periods of time.

If, however, the very considerable sinking of the water-surface, which is shown in the above table to have occurred during the relatively small period of 50 years during low and medium stages, are observed, the sad certainty will be reached that after the expiration of 100 or 200 years the discharge during low and medium stages on the upper and middle portion of these five streams may have been so much decreased that they will no longer be navigable, if the causes which produce this decrease in discharge of streams are not counteracted.

Those creeks and rivers whose basins are small may be converted into torrent streams by this continued decrease in their height of stages and discharge, being probably dry annually during several months, and then being rapidly filled up by rains and made to discharge large quantities of water. There are many examples to show that these apprehensions are well founded. Many large rivers even, which historical researches have proven to have been at one time prolific in water supply, have now become already only wild and torrent streams, as is the case with most of the torrent streams which precipitate themselves down the southern slopes of the Alps in Italy and in Carinthia. Many other creeks and rivers which existed decades ago have been changed only during recent periods within the memory of man into wild creeks which only discharge large quantities of water and débris during heavy rains.

It will therefore seem clear that my apprehensions concerning the future navigability of the five streams above mentioned are well founded.

Although, aside from the data which Dr. Berghaus furnished in his hydrographic work, and which are unfortunately too little known as yet, we have no reliable result of gauge-readings, extending during long periods on most of the large rivers, which would permit tabulation and comparison, many professional men have been led by the data gained by some to observe the decrease in the discharge of other rivers.

We find, for instance, the following interesting communication in the "Deutsche Monatschrift für Handel Schiffahrt und Verkehrswesen, (I Band, Rostock 1872,)" *published by F. Perrot, viz:

It has been proven that the three streams Weser, Elbe, and Oder, clearly show a decrease in their discharge and an increase in the sand deposits in their beds. It has been calculated that if the water supply in the Elbe decreases at its present rate, it will not in the future be navigable by heavily laden vessels. It is not different on the Oder; there were only 11 days during the year 1858, which it is true was quite a dry one, on which the Silesian Oder was navigable under full power. The Weser from the beginning had the least water supply of the three. The government, it is true, has recently taken steps to prevent the clearing of the heights which border the valley of this stream, and which was the main cause of this phenomenon, but the incompleteness of the river improvement, made according to the principles at present in vogue, has done more to bring about the state of affairs which now weigh so heavily.

If the tabulated gauge-readings and other data which have been collected in this treatise, as well as my conclusions, which I believe to be unassailable, are thoroughly criticised, I believe that no hydraulic engineer will doubt the correctness of the assertion already made by the distinguished hydraulic engineer, Dr. Berghaus, in 1836, and proved more in detail by me since that time, that in the creeks, rivers, and streams of Central Europe there have been more frequent and higher floods, and on the contrary a very serious and continuous decrease in the height of low and medium stages, and consequently in discharge, during the period of observation of about 140 years.

A correct knowledge of and heeding of this peculiar phenomenon in the life of rivers and streams is very important for the hydraulic engineer particularly. He must give due consideration to this continuous change in the conditions of discharge of creeks, rivers, and streams in projecting improvements of rivers, canals, aqueducts, water-powers, and all hydraulic works in general which are intended to last during centuries.

If he does not, his works will not be suitable to their intended purposes and will be total or partial failures, as I have already pointed out on the Rhine, Elbe, and Danube.

I flatter myself, therefore, that through the laborious collection and tabulation of the material before us, and through the establishment of the above-mentioned phenomenon, I have contributed my mite to promote hydraulic knowledge, and particularly that of river works.

II. DECREASE IN THE NUMBER OF SPRINGS AND THEIR DISCHARGE.

The theory which is already generally accepted as correct, how rainfalls create and feed the supply of the subterranean water which oozes into and filters through the earth, and how these waters slowly filter through the pervious or so-called water-conducting subterranean layers from the mountain sides and high regions into the low lands and valleys, and there reappear as springs, and in their totality mainly supply creeks, rivers, and streams, is probably well known to my readers. To those who wish to inform themselves more on this subject, I can recommend the Hydrography of Dr. Berghaus, which I have already cited (second volume of his "General knowledge of countries and nations," published in 1837), and the "Handbuch der Wasserbaukunst, von G. Hagen."†

---

* German Monthly for Trade, Navigation, and Commerce, (Volume I, Rostock 1872.)
† Handbook of Hydraulic Art, by G. Hagen.

My esteemed readers will see from these works that by means of abundant and very careful observations, measurements, and calculations it
has been found that only about one-sixth to one-third of the whole rainfall of the basin of a river or stream which reaches the surface of the
earth runs directly on it into the creeks, rivers, and streams, and generally flows rapidly to the sea through them. The amount depends upon
the shape of the surface, the geological formation of the strata, and the
character of the plants which cover it.

The other five-sixths or two-thirds of this rainfall enter the earth, are
gathered in reservoirs, in cracks, hollows, and pervious layers, and supply the water which oozes and filters through the ground as well as the
springs. A clear conception of the extent and magnitude of these subterranean reservoirs can only then be gained when we consider that in
the dry season of the year, during which not a drop of rain falls in the
basin of a river for weeks, all the springs, wells, creeks, and rivers which
are situated in it are fed by these reservoirs alone. They do this so
evenly that although the river stages sink the most, yet there is the
greatest sameness and duration in them.

The great importance of these subterranean waters only seems clear
when we call attention to the fact that mankind draws most of the drinking water for itself and its animals from them, and they may therefore
be considered the fundamental condition upon which the existence of
animal life is based. It is made clearer still when the fact is considered
that the waters which ooze through the upper layer of the earth dissolve
nourishing matter and thus prepare the latter for suction by the roots of
plants, and thus again make the fundamental condition upon which the
existence of vegetable life is based.

The points just indicated show how important a rôle the waters which
ooze into the earth and filter through it, subterraneously, play in the
economy of nature.

I can only give the following proofs of the decrease in these subterranean water supplies:

It has been shown in the previous chapter by observations taken during a long period of years that during the last decades the floods in
creeks, rivers, and streams which are produced by rains occur oftener
and rise higher and therefore now produce larger masses of water in
them than formerly.

It follows, then, that if the amount of rainfall has remained the same,
the amount of water which has flown off on the surface of the earth has
increased, and the quantity which has oozed into it has decreased.

Consequently the mass of the subterranean reservoirs, the seep and
filtration waters and springs must have diminished.

The correctness of this conclusion is confirmed by the following data
which have been gathered.

We have already shown that in the last decades the height of the low
and medium stages and consequent discharge of creeks, rivers, and
streams have decreased continuously. This decrease has been the greatest during those months when these water-courses were almost exclusively fed by these subterranean seep and spring waters. We are therefore justified in concluding that in recent periods the water-supply furnished by these subterraneous reservoirs and pervious layers has decreased, and that the seep and spring waters of basins have furnished
a smaller supply for feeding their water-courses than in former years.

Dr. H. Berghaus has already referred to this matter in his Hydrography (p. 30), as follows:

It has been observed in many regions of the earth that springs lose a part of their water supply. In the former Poitou and in the department of the Lower Charente a striking decrease of springs has been felt since 1825. This phenomenon has been ascribed to the drying up of the soil and to the construction of canals, ditches, &c. Fleurian de Bellevue took pains to show that a decrease in the amount of rainfall was the cause of it.

It has also been established by numerous observations that in recent periods many springs have dried up, and that the discharge of many has decreased considerably. It is also generally known that many aqueducts, which were constructed in a solid manner and at great expense, and which once delivered an abundant supply of water, have been completely abandoned on account of a permanent failure therein. It is further known that very many old wells which were once supposed inexhaustible have, especially since 1852, partly become totally dry or partly so poor in water supply, by reason of the sinking of the surface of the water in the subterranean layers which fed them, that it became necessary to deepen them several fathoms in order again to obtain a permanent water supply.

I will only mention here a few examples of the many of which I have obtained knowledge.

It is generally known that Rome, during its halcyon days, with its numerous public wells and baths, was extravagantly supplied with water by numerous aqueducts. In consequence of the drying up of the springs which fed them, many of the latter are now perfectly dry, and others have but a part of their water supply. An example of the latter is the "Aqua vergine," which is nearly 12½ miles long, and which it is now intended to reconstruct at considerable expense.

The springs and aqueducts which once supplied Constantinople with a great abundance of water for drinking and useful purposes, have now lost so much in their yield that there also new and distant springs must be sought.

The world-renowned fountains and artistic water-works in the palace garden of Versailles were once so abundantly supplied by the aqueducts which furnished the water that they could play during nearly a whole day at a time. Now, in consequence of the decrease in water supply, the latter must be collected during about 23 hours in order that the fountains and works may play an hour.

The many beautifully constructed fountains and artificial waterfalls in the Belvedere, Schwarzenberg, and Lichtenstein gardens at Vienna, and those in the palace park of Schönbrunn, which were abundantly supplied by springs which were led to them, are now completely dry, and are a sad monument of dried-up springs.

The city of Vienna has, in addition to about 10,000 wells, 19 different aqueducts by which the spring and seep waters of the vicinity are caught, gathered, and led into the city. After the water in these wells had not only diminished in quantity but had also deteriorated in quality in consequence of seep from sewers, and the supply formerly furnished by the 19 aqueducts had become greatly diminished, the Emperor Ferdinand aqueduct was built in 1836. This delivered to the city about 1,245,627 gallons* daily from a supply gathered at Nussdorf, from the Vienna Danube Canal, by means of long and deep siphons.

This aqueduct became uncertain and insufficient, partly in consequence of the sinking of the surface of the Vienna Danube Canal, and partly through the clogging of the siphons. The community of Vienna was then induced, after proceedings lasting several years, to draw the requisite supply of water for drinking and useful purposes, amounting to

---

* Eimer (buckets) = 3,454 cubic inches.

24,912,540 gallons daily, from the high springs of the Schneeberg (Kaiserbrunnen and Stixenstein springs). This will be accomplished by an open aqueduct about 56½ miles long, and which will require a capital of nearly $8,000,000.

I believe that I have proven by the foregoing information and examples that during recent periods the supply in subterranean reservoirs and in water-carrying layers has become smaller; that many of the seep waters and springs have completely dried up, and some deliver less water; and finally, mainly for these reasons, the low and medium stages continually sink lower, and that the discharge of these water-courses consequently becomes continually smaller.

If the decrease in water which has been shown to have taken place during the last period of 140 years should continue, this phenomenon and change on the surface of the earth would cause incalculably evil results and dangers to future generations. This decrease and sinking of the water oozed into and filtering through the ground under the surface, and the more numerous changes between very dry and very wet years, which is shown by the plots, would decrease the fertility of the soil greatly, and many countries which are now covered with rank vegetation would become cheerless deserts.

After many springs have dried up, and creeks and rivers have been converted into torrents, mankind would be compelled to draw the water for drinking and useful purposes from deeper water-carrying layers and from greater distances, and consequently at greater expense. Many manufacturing and industrial establishments would also lose their indispensable, useful water and water-power, and would either be compelled to resort to expensive substitutes, or move to distant regions in which the springs and rivers have not yet lost their waters, or stop entirely.

Finally, a continued decrease in the water supply of rivers and streams would lay the former dry during the greatest part of the year, and destroy the navigation of the latter.

Since it is apparent from the foregoing remarks that the continuous decrease of running waters on the earth's surface not only seriously threatens the welfare and health but also the existence of future generations, many friends of natural sciences will probably find themselves incited to make further researches in order to determine the causes of this striking phenomenon, and to ascertain the means by whose application this threatening calamity may be counteracted, so far as it lies within the power of man.

I have been industriously engaged in the solution of this difficult problem, and have compared the results of my investigations on this subject in the following two chapters, in the hope that distinguished colleagues and naturalists may continue the researches thus begun, and make them bring good fruit for the welfare of future generations.

IV. CAUSES OF THE DECREASE IN THE WATER SUPPLY OF SPRINGS AND STREAMS, WHICH HAVE BEEN ESTABLISHED BY THE PRECEDING CHAPTERS.

The accurate determination of the causes which produce the decrease in the water of springs, rivers, and streams, is of the greatest importance, since only when these are known can suitable measures be suggested to set a limit, to the extent of the power of man, to the further progress of the evil.

In order to obtain a correct appreciation of the connection between the amount of rainfall and the discharge of streams, we will present these quantities by an algebraic formula, and for this purpose adopt the following notation, viz:

Let us represent by A the total amount of the annual rainfall in the basin of a stream; by x A that part which runs off the surface into the creeks, rivers, and streams, which will then give for the amount which oozes into the earth $(1 - x)$ A; by B the amount which is consumed by vegetation and animals, as well as that which runs off into very low layers or crevices in the earth and does not reach rivers and streams; and by M the amount of water which is annually discharged into the sea by this stream, and we will have

$$M = x A + (1 - x) A - B = A - B.$$

It will be seen from this equation that the proven fact that x A has now become greater and $(1 - x)$ A less has no influence on M. Now since it has been proven that M has diminished during recent periods, it follows that this decrease could only have occurred in consequence of a decrease in A or an increase in $(1-x)$ A or by a contemporaneous decrease or increase in these quantities.

Now let us represent by m the decrease in the discharge of the stream; by a the supposed decrease in the amount of rainfall, and by b the increase in $(1 - x)$ A, we have the following equation:

$$(M - m) = (A - a) - (B + b.)$$

and therefore

$$m = a + b.$$

From that which will follow it will be seen that in cultivated countries a and b, and hence m, are continually increasing.

The observations and researches which I made during many years led me to conclude that the changes in the conditions of discharge of rivers and streams which have been established, were caused in the first place and chiefly by the clearing and destruction over large areas of the forests which formerly existed in cultivated countries, by which the mountain sides were left bare and the level ground converted into meadows and fields.

The meteorologists and naturalists of the countries referred to had already during many years made numerous observations and researches to show the influence of forests on the amount of rainfall, the condition of the climate, and the fertility of the soil.

I will here briefly mention some of the results obtained and opinions expressed by some of these distinguished experts.

Dr. Berghaus mentions in his Hydrography, heretofore cited by me, that the decrease in the discharge of the Elbe since 1782 is caused by the decrease of the rainfall which has taken place in its basin since that time. The cause of this is stated, both by Berghaus and Malte Brun, to be in the clearing of forests, since the attraction of the latter draws electricity and water from the clouds and increases the amount of rainfall.

The journals of the Austrian Society for Meteorology for 1867, 1869, and 1871 contain many very interesting treatises on the influence of forests on rainfalls and climate, to which I beg leave to refer.

The distinguished naturalist David Milne Home, president of the Scotch Meteorological Society, has communicated the following experiences:

Wood being a poor conductor of heat, it has been established by observations that trees are cooler in summer and warmer in winter than the air in which they grow, and they have therefore the tendency to equalize the temperature of the different seasons. Trees have a cooling effect in summer in three ways: First, they protect the ground against the powerful effect of the sun; second, by the evaporation from their leaves; and third, by the radiation of the latter.

The French naturalist Boussingault, and Strzelecki also, have repeatedly observed while traveling in the tropics that the radiation of forests at night lowers the temperature of the surrounding air to such a degree that whenever the sky is clear the aqueous vapors which float in it are condensed into the form of a shower or heavy dew.

It is a fixed fact, which has been proven by numerous observations, that the clearing of forests dries up springs, and that when they again grow the latter flow more abundantly and regularly. This has been substantiated by the naturalist Becquerel. Boussingault says: "It is my opinion that the clearing of forests over a large area has always the effect of diminishing the annual rainfall." The learned Coultas makes the following comparison:

The ocean, winds, and forests may be considered as the different parts of a great distillery. The sea is the boiler in which steam is created by the heat of the sun. The winds are the pipes which lead the steam to the forests where a lower temperature exists. For this reason the steam is condensed, and in this manner forests distill showers of rain from the masses of clouds floating about in the air.

Similar views have also been expressed by several naturalists, as Hershel, Arago, Känitz, Lecoq, and Tchihateff. The commission which was appointed in England in 1851 to examine this subject arrived at the following conclusion:

In such countries in which the preservation of the water supply is of great importance, the reckless destruction of forests in those regions in which the springs exist must be resolutely advised against.

The learned Blanqui says also:

The terrible droughts which desolate the Cape Verde Islands must be ascribed to the destruction of the forests. On the island of St. Helena, where the wooded portions have increased considerably in area during recent years, it has been observed that the rainfall has been increased in proportion. The amount of this is now double that of the time when Napoleon I resided there. In Lower Egypt, where it rained but seldom in the past century when the French occupied it, and not to exceed twelve days annually, there is now an abundant rain during thirty to forty days in the winter. This change is ascribed to the circumstance that the Viceroys of Egypt have caused about twenty million trees to be planted below Cairo.

The naturalist Marchaud also mentions the following interesting case, viz:

Before the felling of the forests, which has taken place in recent years in the valley of the Sulz, the Sorne delivered a regular and abundant supply of water for the iron-works at Unterwyl, and neither drought nor heavy rains had any marked effect. Now the Sorne has become a wild creek in which every rain produces a flood. After a few days of fine weather it falls so rapidly that it became necessary to change the water-wheels, and finally to erect a steam-engine to prevent a stoppage of the works for want of water.

Meldrum, the director of the observatory on Mauritius Island, has found that since 1852, when at least 44,155 acres* had been cleared, the rainfall, humidity, and pressure of vapor have decreased, and, on the contrary, the floods and dry periods have increased. Dr. Graham is very decided in his opinion that the quantity of water in the springs and rivers of the Madeira Islands has decreased in consequence of the clearing of its forests, and that the trees have a powerful effect on fogs by retaining and condensing them. The report of Mr. Mathieu, professor of the imperial school of forestry, was published by the French Government, in which he shows that it is proven by experiment that the amount of aqueous vapor over a cleared field is five times as great as over ground covered with forests, which he considers equivalent to an increase in the annual rainfall in sections of countries covered with forests. Prof. H. W. Dove, one of the first authorities in the knowledge of weather, says:

3   * Morgen = 3,053 square yards.

Europe has worked itself into continually irregular rainy seasons by the modern cultivation of its lands, which crowds off the forest recklessly, and which causes its rivers to be nearly dry during long periods, while during others their banks can hardly retain the masses of water which are poured into them.

Dr. R. Grüger, an ardent scholar of Dove, asserts in his popular treatise on the knowledge of weather:

In such places where want of judgment or the greed of gain has not disturbed the natural conditions, and cultivated fields and wooded lands alternate in a certain manner, a healthy regulation produces its own rain, which in turn again nourishes it; while on the contrary a thoughtless and planless destruction of the forests, which it is the custom to call cultivation of the soil, often destroys its fertility forever. Sicily, the former granary of Rome, furnishes a fearful voucher for this. Its interior now resembles a desert, not only because, as Liebig supposes, the soil has been exhausted of nourishment for vegetation, but far more because its culture has been driven to the highest point, that is, its mountains have been cleared to their tops.

If the opinions of these naturalists and experts, which are based upon long experience, observations, and experiments, gained and made in different portions of the globe and in different countries, are now added together, it will be found that forests have the following effect on the discharge of springs and streams, the character of the climate, and fertility of the soil, viz:

1. The existence of forests in a country increases the amount of rainfall, because the fogs and clouds, saturated with aqueous vapor, which sweep over it, are, in the first place, condensed by contact with them and fall in the shape of rain. Again, since the temperature in forests is considerably lower in daytime and higher during the night than over the adjacent fields and meadows, a continual circulation of air is created around them, and this draws the fogs and rain-clouds and causes them to empty their contents. It is, therefore, not on account of the forests, but in consequence of the change between wooded and cleared sections that rainfall is increased.* It is also quite probable that forests attract clouds, deprive them of some of their electricity, and thus increase rainfall. It is, besides, a known fact that a large part of rain is held by the leaves of trees, and that a part of this falls to the ground slowly, and a part evaporates and then is again precipitated as fog, dew, or rain. In other words, the same rain is retained longer in wooded countries, is repeatedly precipitated, and increases the rainfall there.

2. Forests increase the amount of the subterranean seepage and water in springs considerably, since the rains, being retained by the leaves of the trees, fall to the earth slowly. They are then retarded in their flow by the spongy surface, and are partly soaked in and partly sink into the deeper layers of the earth. The latter is made more easy by the roots of the trees, which are spread out over a large extent and make rents, splits, and channels in the soil, which cause more water to ooze in and sink deeper in the wooded than in the cleared tracts. It has also been established by experiments that the aqueous vapor which forms over cleared fields is at least four or five times as great as that over a wooded tract, and consequently that the water which has oozed into the soil of the latter does not vaporize so easily, but is retained and serves to feed the seepage springs and rivers.

3. On extensive clearings, especially in mountainous regions, and even in hilly countries, the rain-drops fall upon the bare ground with vehemence, tear it open, and then flow down the mountain-sides with great velocity and carry earth and débris toward the creeks, rivers, and streams. In this manner these water-courses are suddenly filled to excess, and, as the table before us shows, cause higher and more destructive overflows than was the case when the forests existed.

---

* According to Dr. Berger at Frankfort.

4. Extensive clearings in a country also increase the dryness of the air in summer, and this increases the length of dry seasons, which naturally diminishes the fertility of the soil.

These very detrimental effects of clearings show themselves quite plainly in Palestine, Persia, Greece, Spain, and the Canary Islands, which were all once blessed with rich vegetation.

A further cause of the decrease in the discharge of springs and streams of many countries in Europe during the last decades is the emptying of lakes and ponds and the drainage of swamps and marshes.

These reservoirs held large quantities of water during rainy seasons, which would otherwise have passed off in creeks or rivers in destructive floods, and then partly and abundantly fed the subterranean seepage and springs, and through long periods, and partly raised them into the atmosphere as aqueous vapor again, soon to be precipitated in the shape of refreshing showers.

The removal and emptying of these numerous large reservoirs, as well as the construction of the numberless drainage-ditches on roads and in the fields, by which the rain is rapidly led to adjacent creeks and rivers before it has time to ooze into the ground, have also had a very considerable effect in diminishing rainfall, drying up of springs, and in the rapid filling up of rivers and streams during heavy rains. This opinion was presented in the most convincing manner by the distinguished naturalist, Becquerel, in his report to the Academy of Sciences in Paris on the causes of the increasing destructive overflows in France.

The well-known naturalists and professors, Drs. Kerner and Hunfalvy, who are thoroughly acquainted with the conditions of the soil of Hungary, have shown in detail in their treatises on this subject that the improvement of rivers by which it is intended to reclaim about 6.665 square miles of swamps and overflowed lands will surely diminish local evaporation, make the air drier and continental, and finally diminish and lower the subterranean waters. This will result in a sinking of the water surface in the rivers, and both of these experts have therefore pressingly recommended artificial irrigation and afforestation.

A third reason for the proven decrease in the water supply of springs and streams is without doubt the fertilizing, cultivation, and improvement of lands, which has been executed on a grand scale and is increasing from year to year.

It will be clear to all that the sparse grasses of natural pastures, meadows, and fallow fields consume only a small part of the rain which falls upon them, and that therefore the remainder serves to feed subterranean seepage, springs, and rivers. On the contrary, the rank standing grain and forage and juicy garden plants which grow on fertilized soil consume a far greater portion of the rainfall, and consequently leave but a small part for feeding springs and rivers.

Moreover, it is also generally known that in cultivated countries large quantities of water are drawn from the creeks and rivers for extensive irrigation of meadows. Hardly one-half of these ever get back to these water-courses, the remainder being consumed by the forage, which grows in larger quantities and more rankly on these irrigated meadows.

It will be admitted that the amount of water which is consumed by the growing vegetation itself, as well as by irrigation, and thus taken from the streams, is very great. If it is considered that this takes place during nearly the whole of the year in a basin of a stream from 66,650 to 266,600 square miles in extent, although this goes on in separate portions, and apparently in a small amount, yet it amounts to a considerable quantity during the long period of time and over the great area, and has undoubtedly the effect of lowering the mean stages, which we have shown has taken place.

Another circumstance which contributes materially to the making wild and raising of the beds of rivers and streams must be mentioned here. In order to derive the greatest benefit from the soil, the cleared portions and portions of steep mountain-sides are more frequently plowed up and converted into cultivated fields, in order to receive from them even the sparsest yield.

During heavy rains the soil of these fields, as well as the débris under it, are easily washed off, and gutters are made in these plowed mountain-sides by the rapidly rushing waters. These are generally enlarged into deep furrows and gorges, and the whole washed-out mass of earth and stone is swept into the adjacent creeks, rivers, and streams. Here it remains in consequence of the diminished motive power of the water, raises the beds of these water-courses, converts them into wild streams, and assists materially in increasing the height of destructive overflows.

Again, since on these steep mountain-sides very often a layer of earth only a few feet thick lies upon this rock, this clearing and plowing robs it of its cover and protection, and it is loosened. The rains ooze through it to the rock, make the surface of the latter slippery, and then frequently the whole mass of this layer, earth and débris, slides down the steep rock surfaces as so-called land-slides, even into the beds of the creeks, and thus dam the latter and cause the upper valleys to be overflowed and converted into swamps.

Any one who has ever fully observed the ruinous effects of the clearing and plowing of steep mountain-sides, and has followed these effects into the rivers and streams, will surely substantiate my assertion that the destruction and damage caused thereby to the adjacent banks of the lower portions of the creeks, rivers, and streams, and even to the state itself, are often ten thousand times greater than the benefit derived by the land-owner who clears and tills these mountain-sides. If these facts are established by experts, I believe that the government is not only justified, but that it is its duty to forbid land-owners to clear and till their steep mountain-sides, for the protection of adjacent land-owners and stopping the further conversion of rivers and creeks into wild courses; that is to say, from public considerations.

Finally, I must also add to the causes which produce a decrease in the height of stages in rivers and streams the increase of population and domestic animals in cultivated countries. It is generally known that man and his domestic animals consume a large quantity of water for drinking, cooking, and other purposes, which is obtained partly directly from springs, partly from the subterranean water-carrying strata by means of wells, and, finally, from creeks and rivers. The number of domestic animals increases in about the same proportion as the number of inhabitants. Experience shows that man uses for himself and his animals about 16¼ gallons per capita per diem in cities and towns, and that scarcely one-half of this flows back as waste into creeks and rivers.

If I now mention the fact that in the basin of the Danube, about 266,600 square miles in extent, the increase in population during the period of 32 years, in which my tabulated gauge-readings were taken, has amounted to 6,000,000, it follows that the 8,125,000 gallons daily, due to this increased population, which no longer flows back to the Danube, amounts to nearly 400,000,000 cubic feet per annum, and that this must certainly have also had some effect in the large reduction in the mean of the gauge-readings at the station at Orsova.

Now if we consider carefully and collect together in our minds the retroaction which has been shown in this chapter to have been caused in the conditions of discharge of springs and streams by extensive clearings, emptying of numerous lakes and ponds, drainage of marshes and swamps, cultivation and irrigation of fields and increase of population in cultivated countries, we must arrive at the conclusion, independent of the data gained by gauge-readings, that the discharge of the springs and streams of Central Europe must have decreased, and particularly in those portions where the bed of the stream has not been raised by deposits.

The decrease which has been shown in the first chapter to have taken place in the height of the annual low and medium stages, as well as in the discharge of their streams, by gauge-readings taken during long periods, established the correctness of the observations, opinions, and conclusions given in this chapter.

V. PROPOSED WAYS AND MEANS TO COUNTERACT THE CALAMITIES WHICH RESULT ON THE ONE SIDE FROM THE INCREASED DE-STRUCTIONS CAUSED BY HIGHER AND MORE FREQUENT FLOODS, AND ON THE OTHER BY THE CONTINUAL DECREASE IN THE WATER OF SPRINGS AND STREAMS AT LOW AND MEDIUM STAGES.

It has been seen from the detailed proofs given in the preceding chapter that these calamities are not caused by constantly acting natural forces, but are produced mainly by the fact that the inhabitants of cultivated countries selfishly and recklessly drain the earth's surface of its yield and thereby partially change its figure in a disadvantageous manner. Although it is not difficult now to determine measures and arrangements to check these calamities, it is clear that these measures can only be successfully carried out by overcoming many difficulties and the expenditure of considerable means during a long period of years, and only then if, besides the government, the more intelligent at least of its inhabitants recognize the pressing necessity of these measures, and if the former take hold energetically.

It is natural for us, before beginning a great undertaking, to look around in the world and in history to ascertain whether or not some constructions or works have before been executed under similar circumstances and from their success to have our minds placed at ease in regard to the success of the one which we are about to undertake and also to utilize in our own the experiences gained in the construction of the others. I will therefore first mention a highly noteworthy example, and the only one in existence, to show how the oldest cultivated nation of the earth, the Chinese, freed their country from overflows resembling deluges, and converted it into blooming fields.

About the middle of the last century a work of forty volumes appeared, entitled, "Schui-Hing Kin-Kien," or history of the management of Chinese waters, from which some very interesting data were published in the "Allgemeine Bauzeiting" (Wien, 1858,) *from which I will here only briefly mention those which bear upon our theme.

The Chinese Empire, which has at present an area of about 5,554,275 square miles and a population of about 360,000,000, lies on the eastern slope of the Asiatic highlands, and at the same time of the highest mountain peaks of the earth, from which two ranges extend from west to east as far as the sea and divide it into three parts. Many large and powerful rivers and streams rise in these highlands, which are covered with eternal ice and snow, and their floods rise to great heights, which formerly swept over and destroyed the whole country like a deluge.

---

* General Journal of Architecture, Vienna, 1858.

About the year 2300 B. C. a new deluge poured down over China, and Yao, who was emperor at that time, said to the ministers: "The monstrous waters of the deluge have spread and overflowed everything; mountains are buried in its lap; hills are buried therein; its tossing billows seem to threaten heaven; nations are fleeing; who can help them?"

After Minister Kuau, who had been recommended for this purpose, had shown his inefficiency by nine years of unsuccessful work, the emperor called a very intelligent man, Yü Schün, the son of a simple peasant, as joint regent and successor to the throne, and intrusted him with the erection of the works of protection against the water. He executed these with remarkable circumspection and great energy, and yet a part of the good lands were still covered with water after thirty years spent in great exertions and erection of works.

As soon as Schün ascended the throne he called the distinguished man Yü, first as minister and then as joint regent, and he accomplished the most astonishing result during his activity of 62 years.

Yü regulated and canalized the most of the larger rivers; changed their courses; erected strong levees; dug large lakes and ponds many square miles in extent, in order to gather the surplus waters of the floods, to be subsequently used for the irrigation of sterile regions, and he built a whole network of waste and feeder canals, in order to regulate the great velocity of streams, to facilitate navigation, and to furnish the water for necessary irrigation. One of the boldest, greatest, and most successfully executed works of Yü is unquestionably the changing the course of the Hoang-Ho, or yellow river, in upper Schen-Ssi. The mountains of Long-Men formerly diverted this great river to the eastward, and thus often exposed the capital of Ki-tsheu to dangerous overflows. To protect the city against this, and in order to give the course of the stream a better direction, Yü opened a new bed for it through the Long-Men range by blasting out the rocks and transporting the excavated material to a long distance. After the colossal work was completed, and the river had taken possession of its new bed and appeared on the other side of the Long-Men range majestically flowing southward, Yü being satisfied with his work, caused a large inscription to be cut into the rocks, of which, although nearly destroyed by the tooth of time, enough remains after 4,000 years to attest to the enterprising mind and high intelligence of the first minister of Schün.

The foregoing brief reference shows that the two distinguished men, Schün and Yü, can be considered as the real saviors and founders of the Chinese Empire, since with their eminent talents and knowledge of hydraulic constructions, they freed the country of deluging overflows, and taught their people how to repair and prevent damage caused by floods; they had proven to the Chinese that with their dangers they possessed fruitful sources of wealth, and that it was a necessity to maintain and complete the original works, and to fight the rushing waters continually. Although the Chinese have followed this advice faithfully at all times, and have continued their works of protection against water with extraordinary labor and means, yet unusual floods have occurred from time to time; these have destroyed provinces, many cities and towns, and buried hundreds of thousands in their billows, but the tireless industry and the endurance of the Chinese people has always repaired all damage and executed new and stronger protecting works. More than 200 emperors have contributed to the construction of these laborious and artificial works, and which required enormous sums of money, by ordinances, advice, granting marks of distinction, and particularly by very liberal subsidies. Some of the emperors did not disdain making plans and estimates for the new works with their own hands, and the nobles of the Mongolian Tartars who had conquered China in 1280 were compelled to come to the resolution to preserve the hydraulic works and continue their further construction in order to maintain themselves on the throne of China.

I will now enumerate briefly the labor and works which the Chinese did during the long period of 4,000 years, and at an expense of their best powers and tireless industry, for the purpose of making war against, carrying off and making use of the great floods which threatened their country, and the successes which they have gained by these works.

First of all, the streams and rivers of the extensive realm were thoroughly improved; very often led into newly excavated beds and in other directions; canalized and inclosed with powerful levees to prevent their overflow. Many very large lakes and ponds from 844½ to 1,710⅜ square miles in extent were excavated and inclosed with strong dams, in order to catch the enormous flood waters which at times poured down from the highlands, and to distribute them eventually into distant districts for the purposes of making them arable and irrigating them.

In this large country a network of more than four thousand canals, many thousand miles in length, were constructed for the purpose of connecting all the rivers and villages with each other. These had the threefold purpose of taking up the floods of the different rivers; of distributing these and employing them for the irrigation of sterile fields, and to create an extremely lively commerce and the cheapest transportation throughout the whole country for the products of its soil and manufactories.

The greatest and most interesting hydraulic work which was ever constructed by the Chinese is, however, the Yü-Leang-Ho, or as it is generally called, the Yü-Ho, the Emperor's river or Emperor's canal. This canal traverses China from the north to the south without any break; has a length of 1,178½, or, according to some accounts, 2,121 miles; a width according to the requirements of commerce of from 155½ to 1,037 feet, and a least depth of 8 feet 3.56 inches. It passes through mountain ranges which cross its path, crosses valleys, intersects all water-courses flowing from west to east, and therefore the greatest two streams of the realm, the Hoang-Ho and Kiang, and connects the latter. The Emperor's canal, which is justly admired by all, also serves the three purposes, heretofore enumerated, so completely that it has contributed materially to the increase of the agriculture and commerce of the country. In addition to these large canals just mentioned, which form a complete and very intelligently combined system for internal commerce throughout the whole country, there are in China an immeasurable number of smaller drainage and irrigating canals which were constructed for the purpose of draining swampy lowlands, and thus making them arable. In earliest periods, already, so-called artesian wells were dug for the irrigation of fields and particularly large gardens. In the district adjacent to Thibet, 10,000 of such were dug, and the natural fountains of Tsi-Tschiien reach a depth of 3,280.7 feet.

But the success of these works of the industrious and persevering Chinese, which have been carried on uninterruptedly during 4,000 years, is grand and worthy of admiration.

The immense quantities of water which yet pour down from the highlands of Asia, and which formerly deluged the country, are now caught by the many large lakes and canals and applied to making sterile fields fruitful.

The extensive lowlands, which this water once covered entirely or turned into swamps, have been completely drained, cultivated, and planted with mulberry trees, cotton, sugar cane, rice, and all sorts of grain, and now indisputably belong to the most cultivated and fertile lands, in which neither weeds nor wild animals find room any longer.

China is indebted to this for its excellent system of agriculture, the rich yield of its soil, the development in a relative degree of industry in all of its branches, and finally to the facility which has been given to trade and commerce by its improved rivers and numerous canals, for its ability to support its large and dense population. Some of this population are already compelled to construct their places of abode on the streams and canals in order not to divert the soil from cultivation, and thus many large cities have come into existence which, like Venice, are situated in the midst of a labyrinth of canals.

From the foregoing brief reference to the admirable and favorable results of the hydraulic constructions of China we Europeans are assured that we can succeed either in removing entirely, or at least in making harmless, the dangers from our increasing floods, and that we can even make them useful. Our floods are quite small compared with those which formerly deluged China. All that we need is the knowledge, energy, and perseverance of that people in executing the works which are recognized as necessary. It can be seen from the work of M. J. Dumas, entitled "Étude sur les inondations, causes et remède, Paris, Lacroix-Comon, 1857," which received the prize, that it is possible to prevent floods in the rivers and streams of Europe. This book contains a very detailed and thorough discussion of this question. It was published in order to compete for the prize which was offered by the Imperial Academy of Sciences at Bordeaux in consequence of the extraordinary overflows which so terribly devastated and desolated southern France in 1856.

The author, after having studied the subject carefully, says that it is true that it does not lie in the power of man to remove the original causes of floods—that is, the very heavy rains which sometimes occur, and the shape of the earth's surface—but that, however, there are means by which these floods may be made harm'ess.

The author recognizes it as a fixed fact to start with, that floods become more numerous and, so to speak, more regular by the clearing of mountains and cultivation of former wooded areas, and that the meteorological effects produced by clearings are much greater and more injurious than is commonly believed.

As a result of numerous measurements in the four hydrographic basins of the Rhone, Loire, Garonne, and the Seine, Dumas then communicated the very interesting fact that the quantities of waters which pass over the banks of a stream during extraordinary floods, and which produce the overflow, hardly amount to $\frac{2}{6}$ of the annual discharge of the stream, or, on an average, $\frac{1}{510}$ of the total annual rainfall in its basin. Now since the quantity of the waters contributing to overflow is by no means as great as is generally supposed it will be possible to retain this part of the rain, as soon as it falls, in low grounds, reservoirs, canals, ditches, and larger and smaller valleys. The means proposed by Dumas to accomplish this purpose will be considered in stating my final conclusions.

I will now give the following notable examples of means which should be adopted in order to limit the destruction of forests, which is rapidly getting the upper hand, and to facilitate afforestation, especially of mountain slopes.

Colbert, the distinguished French minister of finance, during the reign of Louis XIV, who recognized the incalculable injuries produced by destruction of forests, enacted stringent laws, in 1669, for their protection which were of the greatest benefit to the country.

During the revolution, however, most of the forests were devastated and destroyed, and on the one hand many springs and creeks which were formerly replete with water became dry, and on the other hand the floods of rivers reached an unusual height and caused frequent and great damage.

In April, 1803, already, a new law for the protection of forests was enacted, and as this was soon shown to be insufficient the French government called upon the most competent scientific men, such as Clave, Buissingault, &c., to investigate this important subject. The chambers adopted two laws which were submitted to them in 1859 and 1860 by the Imperial French government, and which were based upon the opinions given by the scientific authorities. The one fixes more stringent limits to cutting down and clearing forests, and the other encourages land-owners to tree-planting. An annual appropriation from the national treasury of one million francs for a period of ten years was made for the latter purpose.

In Scotland also, since 1810, annual prizes are distributed to land-owners for afforestation.

In the United States of America, where the senseless removal and destruction of ancient forests by means of fire has already begun to exhibit great evil effects, the Agricultural Society of Massachusetts voted a prize of $1,000 for that land-owner who should during the next five years plant the largest area with trees.

Mr. David Milne Home, president of the Scotch Meteorological Society, in his work on the increase of the waters in springs on Malta, and the improvement of its climate, gave the Royal English government the following advice, which was based upon his own researches and the observations and opinions of distinguished experts, which have been given by me in the preceding chapter. His advice was that, in order to increase the quantity of water in the springs on Malta, to improve its climate, and to increase its fertility, it was absolutely necessary to afforest the heights on the island. This should be accomplished by the government by first planting the government domains with trees, then by enacting laws for the protection of trees, and finally by offering a premium to every land-owner who should plant a considerable number of acres of land with trees.

After I have thus, in the foregoing, given the opinions and projects of the most distinguished naturalists and experts, and have mentioned the attempts which were made and the experiences which were gained during many years in other countries, it will not now be difficult to propose the arrangements to be made and means to be taken to prevent the great calamities which result on the one side from the more numerous, higher, and consequently more destructive overflows caused by floods, and on the other side by the continual decrease in the discharge of springs, as well as in rivers and streams at their low and medium stages.

I deem it only necessary here to consider these arrangements and means briefly, since their necessity and usefulness is already shown by the opinions and experiences which have been enumerated, and because a detailed description of the manner of making every arrangement would exceed the limit of this treatise.

In order to avoid the calamities which have been mentioned, the following arrangements must be made and measures taken, viz:

1. For the general protection of forests, and particularly against the destruction of these on mountain sides, rational laws must be enacted and their obedience enforced and insisted upon with greater severity. Daily experience teaches that the forestry laws which now exist in most countries are either incomplete or are not obeyed by owners of forests, and that thus one portion of the latter after another disappears.

2. Government should lead with a good example by introducing on its domains a well-regulated administration of forests, and by planting all bare places, especially on mountain sides, with trees. They should also, through well-compiled printed circulars, call the attention of agricultural societies, large land-owners, and separate communities, to the great injuries which result to them through the destruction of forests, and the great benefits that will be gained by them by afforestation. It can only then be expected that the individual communities and land-owners will labor with united strength and assist the efforts of the national government in this direction when the whole population have thoroughly learned the great influence of forests on the fertility of the soil, on the security against overflow, and the maintenance of a regular water-supply in springs, creeks, rivers, and streams.

It is, however, indispensably necessary that the government set aside suitably large sums during a long period of years as premiums for the encouragement and assistance of those land-owners who plant large areas, and especially unfavorable ones, with trees. No real tax should be levied on such afforests during a long period of years, since the owners have in the first years a considerable outlay, and only derive benefit from them after from 30 to 50 years.

3. The cultivation of cleared mountain sides and plowing of steep hill-slopes should be prohibited by law, since from these loosened steep slopes heavy rains carry large masses of earth and débris into the creeks and rivers, create land-slides and dams, fill up and raise the beds of water courses, increase the overflow, and thus imperil the general public interests. Communities and land-owners should also be compelled to stop, without delay, the washing off of hilly slopes, in which gullies and slides have already been formed or are beginning to form. By means of suitable works and plantings, at such points where the evil has already acquired such large dimensions that the execution of such works would exceed the ability of individual communities and land-owners, the government should, in order to promote the general welfare, give suitable assistance.

4. In very steep mountain valleys, in which the creeks at a flood-height undermine the mountain-slopes, roll before them large masses of débris, and then partly deposit these at their exit from the valleys as bowlders, and transport the remainder into the nearest rivers, raise the beds of the latter, and thus give rise to higher overflows, continuous dams must be built across them.

Since these dams hold back the débris in mountain gorges, and detain the waters produced by heavy rains in the basins created by them, and only allow them to pass off gradually, and thus serve a useful purpose, several of them should be built in rear of each other in most mountain gorges and long valleys.

5. The present favorite custom of draining and drying existing lakes and ponds, particularly those which catch flood-waters and sometimes even débris and allow the former to flow off gradually, should not only be prohibited, but the efficiency and value of these reservoirs should be increased by deepening them by re-excavating their muddy and raised bottoms. The expense of this will be repaid in most cases by using the

mud excavated from these lakes and ponds on the adjacent fields, for which it serves as an excellent manure.

6. Large suitable depressions near creeks and rivers which carry off large quantities of water in heavy rains should be converted into reservoirs by inclosing them with strong dams, into which at least a part of the floods can be led, retained, and then allowed to run off gradually. The sides and bottom of these reservoirs can always yet be used as pastures or meadows.

7. From these reservoirs distributing canals and ditches should be led in every direction as far as the shape and conformation of the ground will permit, and thus lead this harmless overflow of water into those regions where they can be employed to the greatest advantage in fertilizing, cultivating, and irrigating the land.

8. At the lowest points of these reservoirs, and in large, level valley-plains, wells, as suggested by Dumas, should be dug. These wells should be dug about 6 feet 6¾ inches in diameter, and down to the lower water-leading layers; then filled with stone, gravel, and sand. Large quantities of rainwater will then ooze into these, and hence they may also be called sinking pits.

The swampy plain of the Paulins at Marseilles, which could not be drained by means of canals, was laid dry by King René by the construction of such sinking pits. If no water-absorbing strata is found at a depth of from 13 feet 3½ inches to 19 feet 10¼ inches, a hole must be bored to such a layer. Mulot bored several of these absorbing artesian wells at Paris to a depth of 265¾ feet, with a diameter of 5.9 inches, which absorbed 130¾ cubic yards of water per hour.

These sinking pits and absorbing artesian wells have also the great advantage in this, that the subterranean layers are fed by them, which increase the supply of water to springs or create new ones at distant lower points.

9. According to the project submitted by Dumas, a whole system of subterranean ditches, 1 foot 7¼¼ inches deep and wide, should be laid out in large, level valley-plains, which should be covered with flat stones, gravel, and earth. The filters or oozing canals expedite the oozing of the water into the upper soil very materially.

10. By the measures which have just been proposed, it is intended to hold back the water which falls during heavy rains in the basin of a stream as much and as long as possible, and thus cause a great part of it to be partially absorbed into the lower strata of the earth's surface, and also to convert a part again into vapor, and then allow the remainder to flow off gradually into the creeks, rivers, and streams, after they have been made useful in the irrigation and cultivation of dry or sterile lands.

But since, notwithstanding these proposed measures, large quantities of water will during long and heavy rains flow into the creeks, rivers, and streams, rise above their banks, and produce destructive overflows, a suitable improvement of those water-courses and a concentration and inclosing of their beds to the normal width is absolutely necessary. In this manner their beds are deepened, their velocity of discharge increased, their water-surface lowered, and the dangers of overflow which formerly existed are in most cases removed.

The improvement or canalization of the Rhine between Hüningen and Mannheim since 1817 gives us a brilliant and encouraging illustration. In many portions of this river, its water-surface has been lowered about 5 feet 2½ inches to 7 feet 3 inches, and the height of flood-stages lowered about 6 feet 2.7 inches, and the former numerous destructive overflows of the beautiful plains in the valley of the Rhine have thus been removed almost entirely.

11. If, notwithstanding the improvement of rivers and streams, and simultaneous confinement of their beds to the normal width, floods still rise over the banks and cause overflow, both banks should be supplied with levees at a suitable distance apart form each other, in order to create a suitably increased cross-section of discharge for floods. But attention must here be called to the fact that the building of levees along rivers and streams before they are improved is very injurious. The beds are then not deepened, but on the contrary filled up with deposits sometimes even higher than the adjacent natural banks outside of the levees, and in case of a crevasse the dangers from overflow become far greater and more extensive. This can be seen on the Po, on most of the rivers in Northern Italy, and on those in Holland.

12. I further consider it my duty to recommend the construction of navigable canals as means for removing overflows in a country, and at the same time promoting its cultivation. By these the floods which collect in the basin of a river during heavy rains can be led into other regions and used for the irrigation of dry lands, and they assist in the evaporation of a part of the water contained in them and in supplying subterranean seepage and spring waters.

Navigable canals are particularly of incalculable benefit in flat agricultural countries, as, for example, Hungary is. The farmer can transport the yield of his land in his own boat to great distances and without almost any expense; the cost of transportation on canals is only about one-fourth of that of railroads, and only about one-tenth of that of hauling on ordinary wagon-roads. They therefore contribute in a very extraordinary degree to the increase in cultivation of the land, the enlivening of internal trade and commerce, and promoting the welfare not only of single speculators, but of the whole populations. This can be plainly seen in Holland and China.

If the measures which I have proposed are carefully considered, it must be admitted that their complete execution in the course of a long series of years will demand great labor, considerable money, and particularly the united and energetic co-operation of all the inhabitants of a country.

If, on the other hand, we look upon the remarkable success and results of these works in China and the plains of the valley of the Rhine, and consider the suggestions made for France in the prize-crowned work of Dumas, we can be confident that we can present the calamities which on one side threaten us from increasing floods of rivers and streams, and on the other from the continual decrease in their discharge during low and medium stages.

It will also be manifest that the execution of these measures will return a large interest, will improve the cultivation and welfare of the countries, and enable the latter to sustain a more numerous population than at present. The emigration of people from Europe to America would then no longer be necessary.

These arrangements and measures, however, can only be inaugurated and begun by the higher governments. They need for their execution the enactment of many important laws, the adoption beforehand of uniform plans of operations for a long series of years, considerable money and labor, and the sympathy of many antagonistic private interests. I therefore take the liberty to commend this treatise to the gracious appreciation of those enlightened statesmen and representatives of the realms who are charged not only with the welfare and prosperity of countries for the present but also for the future.

At the same time I earnestly beg all friends of the natural sciences, and particularly my esteemed colleagues, to subject the highly important question discussed in this treatise to a most thorough examination, and, if convinced of the correctness of my proofs and conclusions, to apply all their power and energy to the timely initiation of suitable preparations and measures to prevent our beautiful mother countries from being gradually converted into deserts, and to guard future generations against severe calamities.

I.—Table of Gauge-readings of the Rhine at Emmerich.

Years.	Mean.		Stage.					
			Lowest.			Highest.		
	Feet.	Inches.	Date.	Feet.	Inches.	Date.	Feet.	Inches.
1770	15	8.86	October 18	9	5.27	December 2	22	10.94
71	13	3.40	November 2	8	5.95	February 7	19	6.78
72	11	7.43	December 11, 12	6	5.23	March 4	20	9.20
73	10	4.80	January 12, 13	5	7.96	January 28	19	6.78
74	12	0.99	December 12, 13	4	10.69	March 3, 4	20	9.20
1775	12	5.41	December 23-25	7	5.59	February 9, 10, 16	20	10.23
76	11	0.22	November 17, 18	5	4.88	February 11	20	8.17
77	10	9.83	October 7, 15, 18	5	2.81	March 2	20	4.05
78	11	1.45	January 14-18	5	7.96	January 29	21	6.47
79	9	3.31	October 30	4	10.69	December 26	21	6.47
1780	10	10.36	January 17, 18	5	4.88	April 8—May 1, 2	19	4.72
81	10	0.76	January 13	5	7.96	January 29	23	2.03
82	10	0.48	December 20	3	8.28	March 28	19	3.18
83	10	8.20	December 26	3	5.70	March 13	22	3.22
84	9	11.96	October 28	5	5.90	March 1	24	7.02
1785	9	4.75	March 6	4	6.58	September 28	18	1.28
86	10	11.81	January 5	4	9.67	February 16	18	10.54
87	9	11.45	January 31—February 5	4	11.73	November 1	19	10.39
88	9	4.76	November 30	4	5.55	March 4	17	4.01
89	13	3.61	January 13	5	5.90	February 3	21	1.68
1790	8	6.46	April 13	5	3.33	December 23	19	5.75
91	9	4.76	October 10	3	11.37	January 19	21	4.47
92	12	7.37	January 16	6	8.32	February 4	23	0.49
93	8	10.58	October 28	3	5.19	February 17	19	1.63
94	8	9.03	January 19	3	6.22	February 19	18	3.34
1795	10	10.78	October 10	4	0.40	February 28	23	8.15
96	9	4.24	December 19	3	5.19	December 26	19	2.66
97	8	9.03	March 31—April 3	3	5.19	July 5	16	5.71
98	8	10.58	November 7	4	10.69	December 9	19	1.63
99	10	10.78	December 22	3	3.49	February 21	25	0.69
1800	7	0.44	August 24	3	1.07	January 17	19	1.63
01	11	3.93	September 15	5	2.81	December 14	22	4.76
1802	9	1.67	October 31—November 6	3	0.04	February 28—March 1	21	2.35
Mean	10	0.74		4	11.42		20	7.55
1803	8	8.00	November 14	3	1.07	February 23	23	7.18
04	11	0.82	October 10	4	2.46	January 4	21	8.53
1805	10	8.72	October 14	3	6.83	February 11	24	5.48
06	11	3.41	November 17	5	2.81	January 25	21	11.67
07	9	5.27	September 7	4	6.55	February 16	22	1.67
08	9	6.30	January 29—Nov. 20	5	5.90	February 6	19	5.75
09	10	6.60	November 24	4	2.46	January 28	24	11.66
1810	8	9.03	February 9	3	5.19	February 30	20	1.99
11	8	4.40	October 27-30	2	9.98	February 3	20	10.23
12	9	3.42	February 2	3	7.25	April 8	20	9.20
13	8	4.30	January 30—February 7	3	0.04	February 22	16	10.86
14	8	9.03	Oct. 25—Nov. 5, 11-13	2	9.98	January 23	24	7.50
1815	8	2.24	January 26	3	1.07	March 28	18	11.57
16	12	8.67	November 4	6	5.23	March 11	20	5.08
17	11	8.15	November 29—Dec. 1	8	0.04	March 13, 14	21	10.59
18	8	6.36	December 31	2	8.95	May 21	19	5.75
19	6	9.35	May 30—June 1	3	8.28	December 29	22	8.88
1820	7	8.78	October 19, 20	4	2.46	January 24	22	11.97
21	10	4.19	Feb. 24, 25—March 2, 3	4	1.43	March 16	19	3.69
22	7	2.19	December 18	3	1.07	January 2	15	11.53
23	8	6.97	Nov. 28, 30—Dec. 1, 2	4	1.43	February 14	17	8.33
24	11	7.43	January 23	5	11.05	November 18, 19	23	6.15
1825	8	7.18	October 22, 23	3	11.37	December 9	20	1.99
26	6	3.58	October 29-31	1	5.51	February 28	14	4.99
27	10	0.48	October 15, 16	3	11.37	March 6	21	8.53
28	8	6.36	November 15	3	6.22	January 18	18	7.45
29	9	3.32	January 18	3	0.04	September 22	16	11.99
1830	9	10.83	December 23-26	5	1.78	February 28	22	10.94
31	11	3.21	February 3	3	6.22	March 10	22	3.73
32	7	0.60	October 31	2	0.71	January 16	19	3.69
33	8	6.15	January 15-17	2	5.86	December 29	23	1.00
34	6	10.77	October 15-20	2	3.80	January 7	23	1.00
1835	6	9.86	December 25	2	2.77	March 21	15	2.26
Mean	9	1.26		3	9.41		20	8.38

II.—Table of Gauge-readings of the Rhine at Cologne.

Years.	Mean.		Lowest.			Highest.		
	Feet.	Inches.	Date.	Feet.	Inches.	Date.	Feet.	Inches.
1782	10	4.60	February 20	3	8.24	March 26	20	1.90
83	10	0.48	December 24	3	7.25	March 10	24	3.42
84	10	6.15	December 31	2	9.98	February 28	41	5.37
1785	8	2.86	January 1	2	5.86	April 24	14	8.80
86	9	7.33	January 7	2	8.95	December 17	18	4.37
87	8	11.61	January 30	3	11.88	October 31	19	11.83
88	8	2.86	December 12	1	5.51	March 2	16	1.86
89	11	11.65	December 16	4	10.73	January 30	23	5.12
1790	7	11.77	April 11	4	0.58	December 21	20	7.14
91	8	9.03	October 8	..	.....	January 17	21	10.60
92	11	8.86	January 10	5	4.88	February 2	25	3.78
93	8	8.00	October 3	3	11.37	February 14	18	6.43
94	..	.....	.....	..	.....	.....	..	.....
1795	..	.....	.....	..	.....	.....	..	.....
96	9	0.64	December 17	3	5.19	December 25	22	1.68
97	8	5.95	March 31—April 1	3	7.25	June 15	17	2.98
98	8	6.97	December 31	3	8.28	December 5	19	4.72
99	10	10.78	December 24	3	6.22	February 24	26	0.01
1800	6	10.38	February 16—March 16	3	2.14	January 5	17	9.16
01	11	2.90	September 15	5	4.85	December 13	23	10.27
02	9	1.15	November 3	2	11.01	March 1	21	9.56
03	8	3.89	November 12	2	9.98	March 5	19	8.49
04	11	3.41	October 10	5	3.85	January 3	23	9.24
1805	10	5.63	February 5	6	3.17	March 7	23	7.18
06	12	0.10	November 17	5	5.90	January 23	24	5.48
07	9	7.33	September 6	4	11.73	March 1	26	6.16
1808	9	4.24	December 29	4	8.64	February 4	21	9.56
Sa.	240	2.27		96	10.76		554	4.27
Mean.	9	7.01		3	11.46		22	2.21
1809	10	3.57	November 23	4	4.52	January 26	27	11.70
1810	9	2.36	October 16	4	5.55	December 20	21	6.44
11	..	.....	.....	..	.....	.....	..	.....
12	..	.....	.....	..	.....	.....	..	.....
13	8	7.90	January 27	2	0.71	February 24	18	0.25
14	7	4.25	October 22—26	3	4.16	March 26, 27	19	9.87
1815	8	10.80	January 29	3	0.29	March 9	22	10.94
16	12	7.89	November 2, 3	6	7.81	January 22	22	10.95
17	11	10.21	December 8, 9	5	10.02	March 12	24	11.66
18	8	10.17	December 30	1	8.60	February 26	18	4.37
19	7	5.90	January 9	2	6.80	December 26	28	3.89
1820	8	2.86	December 31	4	1.43	January 22	27	6.55
21	10	2.85	January 2	3	0.04	March 14	20	5.08
22	7	4.18	December 25	1	1.39	January 1	16	1.80
23	8	9.03	January 2	1	9.62	February 13	19	5.75
24	12	2.95	February 22, 23	5	10.02	November 10	27	10.67
1825	9	1.46	October 20	4	8.64	December 7	22	10.94
26	7	4.15	January 14	2	5.86	February 27	15	2.26
27	10	3.05	February 23	3	8.28	March 4	25	7.90
28	9	3.83	November 10 14	4	5.55	January 17	20	11.26
29	9	6.41	January 25	1	5.51	September 21	17	10.19
1830	9	8.46	January 2	1	9.62	February 28	23	8.12
31	11	11.34	February 3	3	4.16	March 8	25	10.99
32	6	7.91	October 29—November 2	3	1.07	January 15	21	7.50
33	9	6.61	January 15	2	4.13	December 27	26	10.31
34	7	7.96	October 18	2	11.53	January 5	20	7.22
1835	7	9.91	December 24—26	3	1.07	March 20	14	9.12
Sa.	231	0.40		83	5.18		557	11.44
Mean.	9	2.89		3	4.05		22	3.82

III.—Table of Gauge-readings and discharge of the Rhine at Germersheim according to the observations, measurements, and calculations of H. Grebenau.

Years.	Mean Annual Stage in Feet.	Annual Discharge in Cubic Yards.
1840	3.8417	53,630,013,273
41	4.8423	59,489,308,622
42	2.6081	44,504,733,302
43	5.4557	64,456,519,388
44	4.8554	60,386,843,573
1845	4.3338	57,186.116,560
46	4.9342	59,950,733,352
47	3.2905	50,489,422,608
48	2.3392	43,562,866,229
49	2.5294	45,010,291,582
1850	3.7367	52,125,724,853
51	4.0025	55,841,533,109
52	3.2381	49,819,903,973
1853	2.9986	47,622,196,094
Mean for 1840–1853	3.7859	53,154,869,396
1854	1.5911	39,542,033,419
1855	3.6678	52,330,044,304
56	+2.1915	42,806,809,529
57	—0.0023	29,721,831,281
58	+0.5118	33,261,325,584
59	2.0898	41,970,651,432
1860	5.1442	61,720,222,958
61	2.0242	42,694,608,311
62	1.8204	39,707,448,331
63	2.3129	42,705,802,368
64	1.8995	41,402,844,185
1865	0.9514	35,012,984,273
66	3.4054	50,081,554,462
1867	5.4361	64,014,497,076
Mean for 1854–1867	2.3560	44,061,059,828
Decrease in second period	1.4299	9,093,809,569

IV.—Table of Gauge-readings of the Elbe at Magdeburg.

Years.	Mean. Feet.	Inches.	Stage. Lowest. Date.	Feet.	Inches.	Highest. Date.	Feet.	Inches.
1728	7	10.37	January	4	6.57	March 22	14	3.97
29	9	1.97	November	5	0.75	April 12	16	2.62
1730	9	4.21	January	5	1.27	April 7	15	7.93
31	8	8.37	October 19	5	7.96	March 27	16	8.80
32	8	11.33	December	5	7.45	March 2	14	11.18
33	7	4.57	August	5	1.27	December 31	13	3.61
34	9	5.83	November	5	10.54	June 30	15	5.69
1735	9	3.70	December	6	0.08	June 23	16	5.20
36	9	0.95	January	6	4.71	July 26	16	9.32
37	10	7.26	July 20	6	9.86	January 28	16	9.32
38	9	2.83	November	5	11.05	February 18	18	5.20
39	10	3.96	January 2	5	4.36	February 28	15	10.50
1740	9	8.37	January	6	3.87	December 27	16	11.63
41	9	10.68	December 21	6	5.75	January 18	17	3.75
42	9	2.76	December 24	5	3.33	March 17	16	1.50
43	8	11.98	December 24	6	2.14	April 11	13	11.85
44	8	11.14	January 6	4	10.18	March 13	17	2.98
1745	10	1.05	November	6	6.77	March 27	16	11.89
46	7	7.55	June 30	4	6.58	April 30	15	8.78
47	8	7.65	September	5	3.84	December 21	17	0.92
48	10	1.30	January	5	3.84	April 15	17	1.05
49	9	9.69	September 20	6	1.11	February 5	16	0.03
1750	8	7.10	February 5	5	1.27	July 19	18	8.29
51	8	9.35	August 10	5	3.84	March 22	16	10.09
52	8	0.98	July 4	5	7.96	August 9	16	4.42
53	8	4.54	July 13	5	3.84	December 30	16	10.86
54	8	8.25	December 4	5	3.84	January 2	17	2.08
1755	9	7.35	January 3	5	7.96	April 4	17	5.04
56	9	5.87	December 16	5	9.51	April 9	16	7.26
57	8	10.38	July 19	6	2.14	April 7	15	1.24
58	7	7.51	July 5	5	0.75	March 30	16	2.62
59	7	4.82	November 26	4	10.18	January 16	11	4.70
1760	9	1.03	September 9	4	11.95	February 4	16	2.62
61	8	4.71	December 10	4	1.94	February 28	17	3.75
62	8	1.71	December 27	5	0.75	February 25	16	7.26
63	8	3.68	January 19	5	9.94	December 9	13	0.52
64	9	11.78	August 2	6	7.75	January 7	16	5.20
1765	8	5.98	August 11	6	0.59	January 15	14	8.09
66	7	9.52	December 31	5	5.29	July 25	13	5.07
67	8	7.29	January 9	3	7.25	February 20	15	4.33
68	8	5.40	December 20	5	8.48	March 1	16	8.28
69	10	19.31	February 7	6	5.75	December 30	17	1.95
1770	12	5.08	June 23	7	8.16	April 9	17	1.05
71	12	11.49	November 11	8	2.34	July 5	17	6.88
72	9	5.02	December 12	5	8.65	March 6	16	7.26
73	8	7.93	January 6	4	10.39	January 30	13	3.61
74	9	11.16	August 29	5	6.58	March 5	16	2.62
1775	9	3.73	October 2	5	3.84	February 16	17	7.61
76	9	1.03	November 14	5	5.65	February 16	16	5.36
1777	9	3.82	September 1	5	6.58	March 26	16	4.42
Su	457	1.27		278	0.39		802	8.54
Mean for 1728-77	8	10.00		5	6.73		16	1.44
1778	9	3.73	August 31	5	3.85	January 31	14	0.86
79	7	9.27	June 14	4	10.01	December 19	15	7.93
1780	8	6.94	August 14	4	10.34	March 13	17	3.66
81	10	8.22	September 5	5	6.58	February 21	17	6.58
82	8	1.54	September 25	3	10.68	April 9	14	9.46
83	9	8.82	December 24	5	3.85	January 20	17	6.58
84	7	1.47	September 19	4	7.26	March 7	17	5.72
1785	8	7.27	December 26	4	4.52	April 23	18	4.11
86	9	3.21	June 23	4	9.16	August 31—Sept. 1	15	11.53
87	7	6.38	August 20	4	8.32	March 8	12	5.05
88	8	0.62	September 27	5	3.85	April 7	12	5.99
89	9	3.42	September	5	9.51	February 4	17	1.95
1790	6	4.45	August 20	3	5.89	February 19	13	7.22
91	6	5.97	September 27	3	8.79	January 23	10	11.81
92	8	10.04	January 11	3	3.13	December 27	12	2.22
93	7	3.70	January 14	3	10.85	March 10	12	10.72
1794	8	10.69	June 21, July 28, Aug. 2	3	11.63	March 4	16	4.66

IV.—Table of Gauge-readings of the Elbe at Magdeburg—Continued.

Years.	Mean.		Stage.					
			Lowest.			Highest.		
	Feet.	Inches.	Date.	Feet.	Inches.	Date.	Feet.	Inches.
1795	7	9.70	June 14	3	11.63	March 21	13	3.61
96	7	4.51	October 5	4	3.49	December 31	13	3.61
97	6	11.90	August 31	4	1.43	January 1	13	4.43
98	8	2.63	December 15	4	1.43	February 18	14	1.91
99	8	5.83	December 18	4	1.43	February 28	17	7.61
1800	5	7.70	August 19	2	7.57	April 6	12	8.92
01	7	7.02	September 10	4	6.32	March 12	13	1.55
02	7	0.17	October 8	3	8.28	March 30	15	8.78
03	7	8.54	January 27	4	0.40	December 30	15	8.78
04	9	5.92	September 22	5	1.78	January	16	10.86
1805	9	11.04	June	6	4.20	March 4	17	9.50
06	8	9.22	July 26	4	1.43	March 26— September	15	5.00
07	7	7.56	August 30	3	5.19	March 5	17	3.75
08	6	8.95	August 22	3	9.82	April 13	17	7.61
09	6	4.39	September 1-28	3	2.10	February 5	17	10.19
1810	5	4.63	January 15	2	7.23	March 8, 9, 19	16	11.89
11	5	4.44	September 30	1	11.94	February 16	13	5.32
12	7	5.17	February 3	2	10.32	April 9	16	4.68
13	6	1.40	July 8	2	8.26	February 24	15	4.06
14	5	7.66	August 16	2	3.80	March 31—April 1	18	1.28
1815	6	3.30	June 8—November 3	3	4.16	August 17	15	3.30
16	7	6.13	December 14	3	2.30	March 20	18	5.67
17	8	10.01	December 31	3	2.10	March 11	14	10.15
18	5	4.72	December 22	2	7.92	March 28	10	11.81
19	7	2.87	January 1	3	1.07	December 31	14	1.91
1820	5	5.10	August 29	2	6.89	January 31	17	10.19
21	7	7.56	February 28—March 1	4	1.43	March 18	15	10.50
22	6	0.68	July 16	3	5.19	March 25	10	8.05
23	5	7.32	November 2	3	9.31	March 3	11	7.02
24	7	1.46	September 25	4	6.58	July 3	17	0.92
1825	6	5.41	August 10	3	8.28	January 1	11	0.84
26	5	9.30	September 29	3	4.16	May 10	12	9.43
1827	7	3.94	August 13	3	3.13	March 8	17	10.19
Sa	368	0.70		195	11.74		753	7.47
Mean for 1776-1827.	7	4.33		3	11.03		15	0.86
1828	7	10.91	June 25	3	10.85	January 22	16	2.62
29	8	2.20	August 29	4	10.69	December 31	13	8.76
1830	7	10.74	August 3	4	0.91	March 7	18	0.25
31	8	2.86	July 19	5	0.75	March 12	16	4.68
32	5	1.74	November 30	3	1.07	January 20	12	4.28
33	6	6.19	January 3	1	6.35	December 31	16	3.65
34	5	10.70	October 9	2	8.95	January 6	16	10.86
1835	4	3.00	December 13	1	4.48	March 22	8	4.92
36	4	10.69		1	10.65		13	7.73
37	7	4.56		3	3.13		16	6.74
38	7	2.50		3	8.28		17	4.00
39	7	7.65		3	9.31		16	0.56
1840	6	1.11		3	3.13		15	8.44
41	6	7.23		2	9.98		17	5.04
42	4	6.58		1	6.54		13	11.85
43	7	4.56		3	10.34		14	8.09
44	7	11.77		3	10.34		18	0.25
1845	6	6.26		3	3.13		19	1.63
46	6	6.26		3	1.07		18	2.31
47	6	11.41		3	9.31		17	10.19
48	5	4.88		2	7.92		16	10.86
49	6	0.08		2	8.95		12	6.34
1850	7	7.65		3	10.34		18	4.37
51	7	9.71		4	5.55		16	2.62
52	6	7.29		3	5.19		17	4.01
53	6	10.38		2	10.01		16	10.86
54	7	2.50		4	0.40		16	7.77
1855	7	9.71		3	0.04		18	0.25
56	6	2.14		3	3.13		16	3.65
57	4	8.64		2	9.98		8	3.88
58	5	1.78		2	8.95		15	1.24
59	5	1.78		2	2.77		12	6.34
1860	6	9.35		2	2.77		17	8.13

IV.—Table of Gauge-readings of the Elbe at Magdeburg—Concluded.

YEARS.	MEAN.		STAGE.					
			LOWEST.			HIGHEST.		
	Feet.	Inches.	Date.	Feet.	Inches.	Date.	Feet.	Inches.
1861	6	0.08	....	3	1.07	....	13	5.87
62	5	3.84	....	2	5.86	....	18	10.54
63	4	10.69	....	2	6.89	....	10	8.72
64	4	6.58	....	2	0.71	....	10	11.81
1865	4	2.46	....	1	9.62	....	18	0.25
66	4	2.48	....	1	10.65	....	9	11.45
67	7	1.47	....	2	4.83	....	16	3.85
68	6	2.14	....	2	2.77	....	13	5.87
1869	5	6.04	....	2	7.92	....	12	10.46
So.....	258	10.75		130	0.07		644	8.51
Mean for 1824–1869.	6	1.98		3	1.14		15	2.21

V.—Table of Gauge-readings of the Oder at Kustrin.

Years.	Mean.		Stage.					
			Lowest.			Highest.		
	Feet.	Inches.	Date.	Feet.	Inches.	Date.	Feet.	Inches.
1778	5	2.09	July 24-30	3	4.16	March 21	9	3.21
79	5	0.58	May 30	2	2.77	December 26	9	11.87
1780	5	7.94	September 19	2	0.71	March 21	13	7.73
81	4	7.02	September 23	1	5.51	March 27	9	2.70
82	4	3.95	September 30	1	5.51	December 31	8	4.92
83	5	9.10	November 10	2	5.60	January 22	11	1.88
84	5	0.56	October 11	2	3.80	April 30	11	5.99
1785	5	8.86	December 18	3	3.14	April 28	15	6.38
86	6	6.26	July	3	2.10	September 2	11	2.38
87	5	4.57	October	2	5.86	February 25	9	11.97
88	4	10.21	October	2	5.86	April 9	9	11.45
89	6	2.14	June	3	4.93	April 20	10	11.29
1790	3	5.79	August	1	6.02	February 26	7	11.77
91	3	3.16	October	1	1.39	February 27	5	7.96
92	3	2.87	September	1	4.48	March 22	7	8.16
93	4	7.66	September	2	3.29	March 12	7	6.10
94	4	1.55	July	1	8.60	March 10	10	9.75
1795	4	0.60	September	1	10.55	March 2	6	11.41
96	4	2.63	October	2	2.77	January 4	5	11.05
97	3	5.26	August	1	10.65	April 23	6	4.72
98	5	9.12	June	3	3.65	April 11	10	5.11
99	5	4.52	December	2	11.01	March 11	10	5.63
1800	2	11.45	August	1	5.51	April 17	8	3.88
01	4	8.07	February	2	4.83	March 18	8	6.97
02	4	0.29	November	1	9.62	March 22	9	0.12
03	4	3.87	September	2	8.95	December 18	7	6.62
04	6	0.48	September	2	11.01	June 21	10	2.02
1805	5	0.82	August	2	4.83	March 13	11	4.96
1806	5	1.07	July	2	3.29	April 6	8	8.26
Su.	136	0.49		60	4.51		274	4.37
Mean.	4	9.12		2	3.46		9	5.33
1807	4	10.37	September	1	10.65	February 28	10	6.14
08	4	2.48	August	1	11.68	April 23	9	1.15
09	4	3.93	August	2	6.89	February 17	9	0.12
1810	3	4.31	September	0	8.24	March 19	9	11.45
11	2	7.71	September	0	6.18	March 16	7	4.56
12	3	2.07	October	1	3.45	April 9	5	11.05
13	4	9.72	June	1	9.62	September 12, 13	11	5.99
14	3	11.53	November	1	10.65	April 8	13	4.13
1815	3	7.14	November	2	2.77	April 2-8	6	1.11
16	4	11.03	October 24, 25	2	6.86	March 24	9	9.91
17	4	3.93	October 7	1	4.48	March 18	9	1.67
18	3	8.79	July 28, 29	1	11.68	January 31	6	11.41
19	3	11.32	July 25-27	1	7.57	April 14	7	0.95
1820	4	1.36	August 30—September 9	1	4.48	February 12	9	3.21
21	4	5.15	November 21	1	11.68	February 1	7	9.19
22	3	0.90	December 20-22	0	10.30	February 2-4	7	5.59
23	3	0.44	Sept. 20, Oct. 4, 31, Nov. 2	1	2.42	March 5	7	6.62
24	2	11.13	October 5	0	1.03	April 29	6	0.08
1825	2	10.21	Sept 29,30—Oct.2,6,8,23.	0	9.27	January 23	6	2.66
26	3	3.79	October 6	1	1.30	May 15, 16	7	11.77
27	3	10.88	October 19-24	1	1.39	March 18, 19	9	1.15
28	4	7.37	June 25-27	1	3.96	January 27	8	9.03
29	5	8.35	September 18	2	9.98	June 25, 26	9	3.39
1830	4	10.23	August 16	1	4.48	March 20	12	11.49
31	5	8.20	May 31—June 7	2	6.89	September 20, 27	10	0.48
32	3	5.27	November 29	1	4.48	February 16	8	2.86
33	4	3.57	June 28-30—July 1	2	11.68	May 3	10	1.51
34	3	0.36	October 3-8	0	3.09	February 7	10	7.69
1835	1	8.54	September 7	0	3.09	January 13, 14	4	2.46
Su.	112	10.07		42	10.36		251	10.83
Mean.	3	10.69		1	5.73		8	8.23

VI.—*Table of Gauge-readings on the Vistula at Kurzebrack.*

	MEAN.		STAGE. LOWEST.			HIGHEST.		
YEARS.	Feet.	Inches.	Date.	Feet.	Inches.	Date.	Feet.	Inches.
1809	7	3.79	September 3	3	7.25	March 10	17	4.01
1810	5	5.39	October 7	1	10.65	March 16	15	11.53
11	3	9.65	September 27	0	10.30	March 15	12	9.96
12	6	7.03	June 26	3	5.19	August 20	13	4.13
13	8	3.63	June 20	4	1.43	September 3	22	9.91
14	7	3.79	October 27	4	3.23	April 5	17	10.70
1815	8	3.63	June 18	4	11.73	August 7	15	9.47
16	10	0.48	August 24	5	6.93	March 25	21	2.35
17	8	4.40	October 4—November 1	4	4.52	February 4	16	10.86
18	7	4.56	December 20	3	6.98	January 31	20	1.99
19	6	1.84	October 17	2	11.01	February 13	14	5.20
1820	7	8.65	September 6	3	10.34	December 22	15	9.47
21	9	2.44	June 15	5	1.78	March 25	18	7.40
22	6	1.11	June 12	3	0.04	February 14	20	4.05
23	6	0.08	September 26	3	4.68	March 13	17	6.07
24	6	5.84	October 6	3	9.82	January 27	14	3.97
1825	6	7.81	October 21-27	3	8.28	February 12	13	0.01
26	0	1.63	November 4-12	2	10.50	January 2	13	4.64
27	6	1.37	October 21-27	2	3.80	March 12	20	6.11
28	6	6.52	June 27	3	4.68	December 31	20	7.14
29	9	9.05	October 7	4	10.18	April 7	24	0.84
1830	7	5.59	September 9	3	5.70	March 25	20	10.40
31	6	11.41	November 1-10	3	2.10	March 27	16	9.83
32	4	11.73	August 28	3	0.04	March 20	6	9.35
33	5	10.54	July 13	3	4.68	February 22	15	7.42
34	5	6.42	September 27—Oct. 5	1	9.62	February 12	17	9.16
1835	4	1.43	November 19	1	7.05	February 3	8	4.92
36	4	10.69	September 4-6	1	10.65	March 10	13	6.18
37	7	4.82	September 5-10	3	2.10	March 26	19	4.21
38	6	3.43	October 15-19	3	1.07	March 25	19	2.68
39	7	4.82	August 1	3	5.19	March 31	20	3.02
1840	9	1.67	August 4	5	2.81	February 2	16	6.74
Su.	219	11.54		109	1.84		542	0.67
Mean.	6	10.64		3	5.19		16	11.38
1841	6	1.11	November 10	1	5.51	March 25	21	2.86
42	3	0.56	September 21	—0	4.63	March 12	8	3.88
43	5	10.02	May 26	1	6.02	February 2	16	1.50
44	9	8.65	July 6	4	0.91	April 2	22	1.07
1845	8	1.31	October 2	3	11.37	April 8	22	2.70
46	7	0.45	November 24	1	3.45	March 8	20	10.23
47	5	4.10	June 12	2	3.80	June 23	10	9.75
48	4	2.20	September 23	0	5.15	March 2	14	0.84
49	5	0.76	August 24	1	5.51	February 4	19	4.72
1850	6	11.41	September 7	1	6.54	March 15	18	9.52
51	6	2.91	October 29—November 1	3	7.25	March 23	16	9.83
52	4	7.61	October 15-18	0	5.15	April 9	12	9.43
53	8	3.12	January 12	0	4.72	May 3	17	9.16
54	7	8.65	November 24	3	2.10	March 18	21	2.54
1855	10	0.99	December 2	3	4.16	March 27	28	3.81
56	5	3.85	November 4-8	1	0.36	February 19	13	11.85
57	4	8.64	December 1	0	5.15	April 10	13	5.07
58	4	1.17	November 16	—0	6.18	April 1	13	6.70
59	3	3.90	August 26	—0	5.15	March 14	12	4.28
1860	6	7.55	November 14	1	8.59	April 10	19	8.87
61	4	5.63	November 4	0	2.07	February 21	19	0.60
62	2	6.60	November 29	1	3.45	March 29	13	6.50
63	0	4.12	September 1	—2	0.71	January 21	6	3.17
64	1	1.43	January 6	0	0.00	July 13	10	10.78
1865	4	1.69	August 3	0	6.18	April 18	18	3.84
66	2	6.38	December 2	—0	5.15	April 1	9	5.27
67	7	4.56	October 3	1	10.65	July 19	20	7.14
68	5	9.76	November 28	—0	8.24	March 6	20	1.99
69	3	3.13	September 22	—0	2.07	March 25	11	5.99
1870	6	0.08	September 18	0	7.21	April 8	21	1.31
1871	6	4.20	September 24	0	10.30	March 3	21	8.53
Su.	169	0.61		32	9.36		520	8.52
Mean.	5	5.39		1	0.70		16	9.14

VII.—Table of Gauge-readings of the Danube at Vienna.

Years.	Mean.		Stage.					
			Lowest.			Highest.		
	Feet.	Inches.	Date.	Feet.	Inches.	Date.	Feet.	Inches.
1826	1	9.97	January 17	—1	10.82	June 21	7	10.41
27	3	6.02	February 17	—0	8.30	June 11–13	8	6.71
28	3	3.94	December 10, 19, 12	0	3.11	September 18	8	2.56
29	3	4.20	January 23	—0	10.37	June 12	9	1.97
1830	3	0.31	December 29	—1	8.67	September 26	7	0.36
31	2	5.83	February 3	—3	0.31	March 7	8	1.52
32	0	3.11	November 1	—2	0.90	August 1	7	7.30
33	1	11.60	January 10	—3	5.50	August 5, 6	8	7.75
34	1	2.26	December 31	—2	10.24	January 2	8	3.60
1835	1	0.45	December 28	—2	8.16	September 17	6	7.89
36	1	7.45	January 3	—2	7.13	February 8	7	0.36
37	1	4.08	February 12	—2	8.16	June 23	7	11.45
38	1	6.67	December 27	—3	10.69	June 26	6	8.81
39	1	2.78	January 6	—3	0.27	May 31	7	11.45
1840	1	3.04	December 20	—4	8.02	August 2	9	3.01
41	1	6.93	November 21–23	—2	9.20	February 25, 26	7	3.15
42	—0	6.74	January 15–17	—4	8.02	April 3	5	5.36
43	1	7.97	January 27	—2	5.05	July 5	7	10.41
44	1	0.10	December 30	—4	10.10	May 1	6	2.70
1845	0	8.82	February 15	—6	3.74	April 3	9	1.97
46	1	6.68	December 19	—2	10.24	February 3	8	6.71
47	1	10.56	January 17	—4	9.06	May 3	8	0.49
1848	0	9.34	December 28	—4	4.91	February 11	8	2.56
Mean	1	7.97		—3	0.31		7	9.38
1849	1	7.45	January 1	—3	3.42	August 25	8	9.82
1850	2	7.64	January 23	—2	6.08	June 22	6	6.85
51	2	6.86	February 28	—2	0.90	August 7	8	2.56
52	1	10.82	January 10	—1	7.71	August 26	7	2.11
53	1	3.04	December 28	—4	2.84	June 22	9	5.09
54	1	4.08	October 5	—1	10.82	February 9	5	8.48
1855	1	8.75	February 22	—3	0.31	August 16	7	11.45
56	0	10.89	November 22	—3	3.42	June 29	7	10.41
57	—0	11.93	January 12—Dec. 22	—4	3.87	June 3	4	9.06
58	0	5.44	January 8	—4	9.06	August 4	7	10.41
59	0	9.34	January 15	—3	10.69	March 8	6	8.93
1860	1	4.08	February 26	—3	1.35	January 3	6	4.77
61	0	6.25	December 30	—4	3.87	June 13	8	6.71
62	0	6.23	January 9	—4	10.10	February 5	12	2.29
63	—0	0.78	December 9	—3	5.56	June 24	5	3.29
64	0	9.08	December 27	—5	1.21	July 15	6	6.85
1865	—1	1.75	December 30	—6	2.70	April 13	6	0.62
66	—0	6.23	January 2	—5	8.48	August 9	4	8.02
67	2	1.42	December 12	—2	4.01	May 4	8	2.56
68	0	9.59	November 25	—3	2.39	May 4	7	2.26
69	0	5.44	January 25	—3	3.42	December 22	5	9.51
1870	0	8.30	February 21	—3	5.50	November 4	6	8.93
1871	0	6.48	December 10	—5	3.29	January 21	10	11.76
Mean	0	10.89		—3	5.50		6	11.00

VIII.—Table for the comparison of Gauge-readings on the Danube at several stations.

In accordance with an edict issued September 9, 1853, No. 7,062, by the Ministry of Commerce, the zero of gauge was changed as follows (+ raised, — lowered,) in order that the zero stage at Vienna should be at zero on all the other stations:

YEAR.	DAY AND MONTH.	VIENNA Tabor Bridge.	NIEDER WALLSEE.	GREIN.	STRUDEN.	MÖLK.	STEIN.	TULLN.	NUSSDORF.	VIENNA Danube Canal, Ferdinand Bridge.	FISCHAMEND.	HAINBURG.
1853		c 0.00	−1 6.67	+2 0.90	+1 10.82	+1 1.49	+0 2.07	−0 2.07	−0 6.23	+0 8.30	−1 8.75	−1 2.52

In the following years the following successive changes occurred in the readings of the other Danube gauges for the zero stage at Vienna, in consequence of changes in the bed:

YEAR.	DAY AND MONTH.	VIENNA Tabor Bridge.	NIEDER WALLSEE.	GREIN.	STRUDEN.	MÖLK.	STEIN.	TULLN.	NUSSDORF.	VIENNA Danube Canal, Ferdinand Bridge.	FISCHAMEND.	HAINBURG.
1855	October 4	0 0.00	−1 1.04	0 9.34	0 3.11	0 6.23	0 7.20	0 5.19	0 3.11	0 11.41	0 5.19	0 10.37
56	September 17	0 0.00	−1 1.04	0 11.41	0 6.23	0 8.30	0 5.19	0 7.26	0 4.15	0 10.37	0 6.23	1 6.68
57	October 4	0 0.00	0 2.67	0 4.60	0 5.19	1 7.26	1 6.23	1 0.45	0 3.11	0 5.19	1 7.26	1 6.63
58	September 25	0 0.00	0 6.23	1 0.45	0 8.30	0 10.37	0 8.30	0 8.30	0 6.23	0 6.23	1 1.49	1 6.48
59	November 2	0 0.04	0 0.45	1 8.73	0 6.23	1 7.26	1 7.26	1 0.45	0 5.19	0 3.11	1 3.14	1 3.56
1860	November 30	0 0.00	0 5.19	1 11.46	1 4.60	1 2.52	0 10.37	1 4.15	1 10.37	0 6.34	0 11.41	1 3.56
61	February 11	0 0.00	0 2.47	1 10.82	1 4.60	1 0.43	1 0.43	1 9.34	0 0.45	0 0.45	0 4.61	1 0.45
62	September 23	0 0.00	0 4.15	2 1.94	2 8.75	1 2.52	1 4.60	1 3.11	1 6.23	0 10.37	2 10.82	1 6.08
63	September 3	0 0.00	0 5.19	1 4.75	1 0.29	1 3.64	1 8.68	1 2.07	1 6.23	1 2.97	1 4.60	1 7.71
64	October 4	0 0.00	1 3.11	1 1.21	1 6.64	1 0.45	1 0.45	0 4.15	1 4.15	1 10.37	2 4.60	1 4.60
1865	September 11	0 0.00	0 3.13	5 1.21	1 6.64	1 1.49	1 3.56	0 3.11	1 3.11	1 6.23	2 11.27	1 3.56
66	September 9	0 0.00	0 2.07	4 4.91	4 0.76	1 3.56	1 3.56	0 0.40	1 5.19	1 3.56	1 11.46	1 10.82
67	August 15	0 0.00				1 3.56	1 3.56	1 0.40	0 0.09	0 10.37	1 11.46	2 2.82
68	September 1	0 0.04				—	1 2.62	0 3.11	0 8.30	1 8.30	2 1.46	1 3.56
69	August 29	0 0.00				—	1 6.64	1 3.11	1 4.15	1 2.52	2 9.29	2 10.82
1870	September 28	0 0.00				—	1 3.56	1 3.11	0 4.15	1 2.52	2 9.29	2 1.94
1871	September 1	0 0.00				—	1 3.56	1 4.15	0 4.15	1 4.60	3 7.58	2 2.07

FLOOD STAGES.

YEAR.	DAY AND MONTH.	VIENNA Tabor Bridge.	NIEDER WALLSEE.	GREIN.	STRUDEN.	MÖLK.	STEIN.	TULLN.	NUSSDORF.	VIENNA Danube Canal, Ferdinand Bridge.	FISCHAMEND.	HAINBURG.
1855	August 15, 16	7 11.45	12 8.51	30 9.00	20 7.98	12 3.32	10 10.73	K 10.86	10 4.50	10 1.39	10 11.76	9 11.18
1856	June 28, 29	7 10.41	11 6.44	30 2.77	20 4.65	12 3.10	11 6.67	K 9.83	9 10.27	9 8.20	11 10.73	11 11.18
1862	February 3, 4	12 2.29	22 9.90	40 6.56	33 1.36	24 7.69	19 4.40	14 3.19	16 6.16	16 9.80	17 0.39	20 2.78

IX.—Table of Gauge-readings of the Danube at Orsova.

Years.	Mean.		Stage.					
			Lowest.			Highest.		
	Feet.	Inches.	Date.	Feet.	Inches.	Date.	Feet.	Inches.
1840	8	8.53	January 20-22	3	5.50	December 1	12	8.51
41	8	10.08	October 1	3	4.46	April 14	15	10.90
42	8	0.23	January 27	1	1.49	April 23	13	10.00
43	10	4.50	October 23	4	11.14	February 27	13	0.66
44	11	1.84	February 6	2	6.09	March 24-30	15	7.79
1845	11	7.02	February 23	3	2.39	April 30	19	7.51
46	10	4.50	November 28, 29	3	6.54	April 27	16	8.82
47	10	8.01	December 31	3	8.61	May 12-18	15	7.79
48	8	5.42	January 26	1	2.52	March 22-24	16	3.05
49	8	2.04	January 17-19	0	6.23	May 4	13	2.74
1850	10	11.76	January 10-11	2	11.27	May 19-22	16	8.16
51	11	1.32	March 13	3	2.39	December 4, 5	15	4.67
52	10	1.64	September 25	6	2.70	April 28, 29	13	7.02
53	11	4.95	December 16	2	1.94	May 7	19	6.47
54	7	10.15	September 28, 29	3	0.31	May 26, 27	11	4.95
1855	10	1.90	December 22	—0	6.23	April 20	17	7.65
Mean for 1840-'55	9	9.75		—0	6.23		19	7.51
1856	8	0.49	November 12, 13	1	5.64	February 28	13	3.77
57	6	9.44	December 28-31	1	8.75	April 3-6	13	0.66
58	7	1.85	January 15, 16	—1	4.60	April 15-17	13	3.77
59	8	7.23	January 20	2	10.24	June 2-5	14	4.22
1860	11	0.28	November 14	4	8.03	April 29	17	9.72
61	8	0.49	November 3	0	9.34	April 7	13	4.84
62	6	8.66	January 11	—0	9.34	March 10	15	3.64
63	5	2.25	September 27, 28	1	7.72	May 16-21	9	3.01
64	8	9.56	January 10	—1	0.45	June 29	15	1.50
1865	7	2.39	December 30	0	3.11	May 3-5	16	11.35
66	5	11.33	January 11, 12	—1	10.83	April 1-5	10	4.50
67	9	9.24	September 22	3	3.42	May 13, 14	17	2.46
68	8	6.97	October 24	2	7.13	May 25-27	16	3.05
69	7	10.15	October 20	2	4.01	December 31	13	11.04
1870	10	11.24	February 11	3	10.69	November 29	15	7.79
1871	10	11.24	September 30—October 1	3	0.31	June 5-7	15	11.95
Mean for 1856-'71	8	3.60		—1	10.84		17	9.72

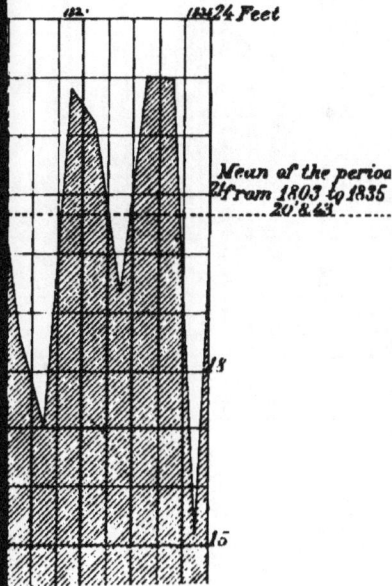

Sheet 1.

Mean of the period
from 1803 to 1835
20'. 63.

24 Feet

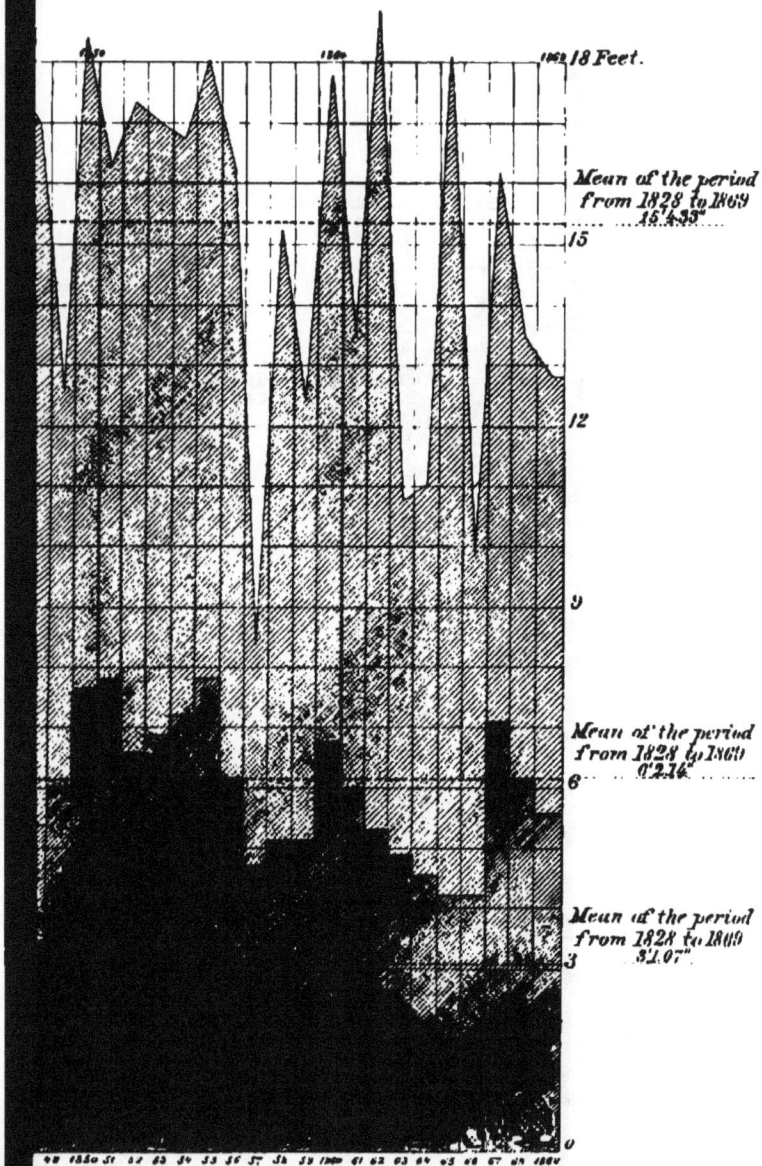

18 Feet.

Mean of the period
from 1828 to 1869
15'4.55"

15

12

9

Mean of the period
from 1828 to 1869
6'2.76"

6

Mean of the period
from 1828 to 1869
3'1.07"

3

0

48 1850 51 52 53 54 55 56 57 58 59 1860 61 62 63 64 65 66 67 68 1869

Sheet 5.

he Vistula

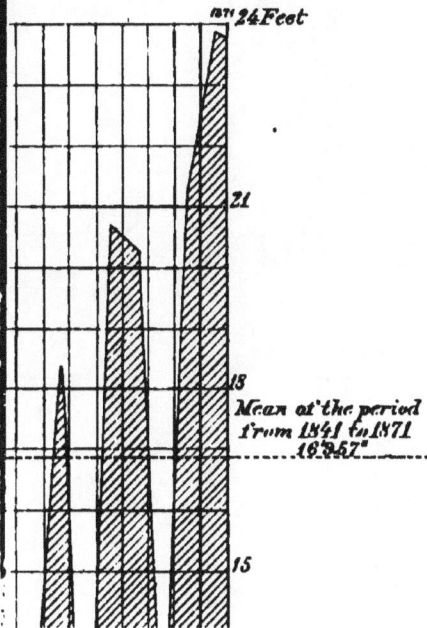

24 Feet

21

18

Mean of the period
from 1841 to 1871
18.957"

15

SECOND TREATISE

ON THE

DECREASE OF WATER

IN

SPRINGS, CREEKS, AND RIVERS,

CONTEMPORANEOUSLY WITH

AN INCREASE IN HEIGHT OF FLOODS IN CULTIVATED COUNTRIES,

BY

SIR GUSTAV WEX,

IMPERIAL AND ROYAL MINISTERIAL COUNSELLOR AND CHIEF ENGINEER
OF THE IMPROVEMENT OF THE DANUBE, AT VIENNA,

WITH SIX SHEETS OF DRAWINGS

FROM THE PAPERS OF THE SOCIETY OF THE AUSTRIAN ENGINEERS AND ARCHITECTS, 1879—Nos. 6-9.

TRANSLATED BY

G. WEITZEL,

Major of Engineers, Brevet Maj. Gen. U. S. A.

WASHINGTON:
GOVERNMENT PRINTING OFFICE.
1880.

SECOND TREATISE

DECREASE OF WATER IN SPRINGS, CREEKS, AND RIVERS.

OFFICE OF THE CHIEF OF ENGINEERS,
UNITED STATES ARMY,
Washington, D. C., July 1, 1880.

SIR: An interesting paper on hydraulic engineering entitled "Zweite Abhandlung über die Wasserabnahme in den Quellen, Flüssen und Strömen bei gleichzeitiger Steigerung der Hochwässer in den Culturländern" (Second treatise on the decrease of water in springs, creeks, and rivers, contemporaneously with an increase in height of floods in cultivated countries), by Sir Gustav Wex, imperial and royal ministerial counsellor and chief engineer of the improvement of the Danube at Vienna, has recently been translated, at my request, by Bvt. Maj. Gen. G. Weitzel, United States Army.

As this paper contains information of value to the officers of the Corps, I have the honor to recommend that it be printed, for the use of the Engineer Department, at the Government Printing Office, and that 500 copies be furnished on the usual requisition.

Very respectfully, your obedient servant,

H. G. WRIGHT,
Chief of Engineers, Brig. & Bvt. Maj. Gen.

Hon. ALEXANDER RAMSEY,
Secretary of War.

Approved:
By order of the Secretary of War.

H. T. CROSBY,
Chief Clerk.

JULY 6, 1880.

VIENNA, *October 12, 1879.*

GENERAL: I take the liberty of transmitting to you in the accompanying package my second treatise, entitled "On the decrease of water in springs, rivers, and streams, simultaneous with the increase of floods in cultivated countries," because I am convinced that such a distinguished hydraulic engineer will take a lively interest in the definite solution of a hydraulic question which is of the highest importance to coming generations and future condition of cultivated countries.

At the same time you will please have the kindness to accept this copy as a mark of my most eminent esteem.

Finally, I take the liberty of politely begging you to transmit for use the other copy of my treatise which accompanies this to the esteemed societies of engineers which exist in your country, with the request that they may thoroughly examine my theory with regard to the decrease in water supply, and illustrate it more clearly by the publication of the hydraulic observations made on American rivers and streams.

With the assurance of my highest esteem, I remain your devoted

SIR GUSTAV WEX,
Imperial and Royal Counsellor.

Brig. Gen. A. A. HUMPHREYS,
Chief of Engineers, Washington.

INTRODUCTION.

In my first treatise bearing the foregoing title, published in the papers of the Society of Austrian Engineers and Architects, 1873, after observing gauges during many years at nine different stations on the five main rivers of Central Europe, viz, the Danube, Rhine, Elba, Vistula, and Oder, and also after many other observed results, I furnished the proofs that the discharge has diminished considerably during the latter decade, not only in the above-mentioned rivers themselves, but also in their tributary rivers, creeks, and springs.

In that treatise I also thoroughly discussed the causes of this peculiar hydraulic phenomenon, and finally proposed measures and precautions to prevent as far as possible the further spread of this calamity, so threatening to future generations.

Since the publication of that treatise, this hydraulic question thus raised by me has been industriously studied in nearly all countries, partly by scientific institutes and partly by individual naturalists and experts, and their various opinions and views have been published.

In by far the greater number of the opinions which have thus appeared some of these scientific institutions and authors have entirely, and others only partially, agreed with my views, conclusions, and the theory which I recommend, and, at the same time, have brought forward many new illustrations and arguments to establish its correctness.

But especially, as several engineers and hydraulic experts have expressed the opinion that my conclusions as to the decrease in the height of water in rivers are unreliable, and that the remaining proofs for the decrease in discharge are not quite unassailable, and, besides, because several of these experts have even advanced the new theory that the established decrease in the height of the water is not in consequence of the decreased discharge, but rather due to changes in and particularly to the deepening of the river beds, I have during six years shunned neither labor nor expense in obtaining as many and reliable technical hydraulic measurements and data as possible of different streams, so that I might clearly ascertain whether the views previously expressed by me or those of my opponents were more nearly correct.

Now that I have in the course of these uninterrupted and continued collections of data and studies, and even by the latest of these, arrived at the fullest conviction that the theory advanced by me relative to the decrease in the discharge of springs, rivers, and streams has been unfortunately proven true, I feel myself impelled to publish this second treatise, since I believe that an indisputably clear exposition of this water question is not only of great importance in technical hydraulics, but also in its bearing upon questions in natural science and history of culture, as I have already pointed out in more detail in my first treatise. But a general knowledge of the decrease in discharge of springs and rivers and the resulting highly detrimental consequences for the future cultivation of countries and their inhabitants is absolutely necessary,

for then alone it can be hoped that the general governments, corpora‐
tions, large land-owners, and individual communities will take suitable
precautions and measures to confine the calamity to such limits as are
yet feasible.

In Chapter I of this treatise I shall present, very briefly, the opinions
given on the question by scientific institutes first, and then those of the
individual naturalists and experts who agree with me; and subsequently,
in Chapter II, with the latest data collected by me as a basis, will fur‐
nish the proof that the objections made by my opponents, and the new
theories established by them, are either groundless or are based upon
incorrect premises.

CHAPTER I.

1. The Imperial Academy of Sciences in Vienna, at my request, ap‐
pointed from its own members a commission of five experts to give a
formal opinion upon my treatise, which commission, after a thorough
examination, in their report dated April 23, 1874, declared themselves
in general as agreeing with me in my proofs and conclusions, and only
remarked that my conjectures that the rainfall in cultivated countries
must now have equally decreased, as not sustained by the meteorolog‐
ical observations made in England, Scotland, and at Paris during a
period of many years, and although it would be hasty to conclude from
this that no such changes have taken place in the interior of the continent,
yet the measurements of rainfall at the continental stations (which, how‐
ever, do not reach so far back) are not favorable to such an assumption.

The Imperial Academy of Sciences at its anniversary meeting of May
30, 1874, accepted the above-mentioned report of its commission, and
resolved at the same time:

(a) To call the attention of the Imperial and Royal government to the
continued decrease of discharge in springs and streams, as well as to the
causes of the phenomenon, and at the same time to apply to the higher
ministries, so that the measures and precautions proposed by Wex to
prevent, as far as feasible, the further spread of this calamity may be
thoroughly considered and carried into execution by the enactment of
suitable laws.

(b) To send a copy of Wex's treatise to the scientific institutes of for‐
eign countries with the request that they furnish the Imperial Academy
with the results of observations extending through a series of years of
the gauges in their rivers, and to furnish this information arranged and
plotted in a manner similar to that adopted in his treatise.

(c) To express, also, to the Austrian Government the desire that sta‐
tions may be erected at suitable points on several rivers, not only to ob‐
serve the stages, but also to measure the discharge, in order to examine
more closely the laws of dependence of the latter on the former.

For these resolutions, which were prepared by the Imperial Academy
and carried into execution, I must before proceeding further return my
sincere and grateful thanks to it, for thereby the scientific societies in
other countries were incited to cause accurate observations and study
of this important water question to be made.

2 and 3. As a result of this the Imperial and Royal Academies of
Sciences of St. Petersburg and Copenhagen appointed their own com‐
missions to examine my before-mentioned treatise, which in their reports
on this subject, dated respectively January 22, 1876, and October 7,
1875, declare themselves as generally agreeing with me in my views and

conclusions, but at the same time remark that a diminution in the rainfall was not established for those localities by the meteorological observations taken through a long series of years. The St. Petersburg commission added, however, to its opinion " that it might, nevertheless, be possible that a difference in distribution of the rainfall and of the evaporation from the surface of the earth might cause a decrease in the total amount of water in some and an increase of the same in other rivers."

The same commission, referring to their observations on the results of clearing lands, also mentions that in the southern parts of Russia, where notoriously one hundred and fifty or two hundred years ago there still existed large wooded tracts, at the present time bleak steppes have developed themselves, whose elevated portions are entirely destitute of water, and whose inhabitants are compelled, therefore, to settle on the banks of the meager and deficient rivulets in order not to perish from want of water.

The same commission still further adds that on the Lower Volga and Dnieper the neighborhood is cleared, and that in consequence the river beds fill with sand-bars, change, and become shallower, but that the floods rise higher than formerly.

The commission of Copenhagen expresses in its reports the following noteworthy opinions :

"Since in the later times our forests have not undergone any change, it is difficult to substantiate the gradual decrease of water in consequence of clearings, but we must nevertheless add that everything leads to the conclusion that at the time when Denmark was covered with large forests, the volume of water in our rivers must have been far greater than it is now, for there are many rivers in the country which exhibit traces of having had at one time a considerable discharge, but now have only a very small one or none at all.

"That the change of woods into cultivated fields has caused a considerable decrease in the volume of water in Danish rivers, the commission can prove as an indisputable fact by the lakes near Copenhagen.

"These lakes are situated in an extensive and flat country, the soil of which consists of clay overlying a stratum of chalk. This entire plain was once a forest, which has gradually partially disappeared, so that some of these lakes are now found in the open fields, and the others in the woods.

"The observations which were made have now established that those lakes which are situated in or near woods receive a much greater volume of water than those of the same area which are situated in cleared country, and as the other circumstances are precisely the same one must conclude that the decrease in the volume of water in the latter is to be ascribed to the clearing of the woods."

The Danish commission finally expressed its opinion " That the destruction of forests results in a considerable decrease in the discharge of rivers, and particularly in that of springs, and, furthermore, that the increased cultivation of fields, and also their artificial irrigation, cause a still greater decrease in the volume of water in springs and rivers."

4. The Royal Board of Canal Directors in Norway intrusted Lieut. Hans Nysom with the task of collecting the data desired by the Imperial Academy of Sciences of Vienna, who, in compliance therewith, in his report dated May 20, 1877, shows that at the two water-gauge stations at Nastangen and Sarpfos, on the Glommen River, which have until now not been affected by any improvement of the river, the observations taken during thirty years confirm the theory of Mr. Wex in regard

8

to decrease in the volume of water in rivers, as well as his opinion of the causes of this phenomenon.

5. At my request the Society of Austrian Engineers and Architects also appointed a committee of experts to give an opinion on my treatise, which, in its report on this subject, made April 17, 1875, expressed the following opinions, viz:

(a) The committee recognizes the vast importance of the question raised by Mr. Wex, but, at the same time, the difficulties of a reliable reply thereto from the observations and data which so far have been laid before it.

(b) In the opinion of the committee it cannot safely be concluded that because there is a decrease in the mean stage there is a decrease in the discharge of a river, because if there should be a change in the cross-section of discharge, or in the slope of the river, the gauges are no longer reliable, and that it is, therefore, absolutely necessary that accurate and regular hydraulic measurements be taken at a number of constant cross-sections on the main river and its tributaries.

(c) Although the accurate measure of decrease in volume of water, expressed in figures, cannot be determined from the data furnished by Court Counselor Wex, yet they point to the fact that in the rivers mentioned there has been a decrease in the volume of water.

(d) The data furnished in the treatise show clearly the fact, of so much greater importance to the engineer, that the regimen of the rivers mentioned has lately undergone a considerable change.

(e) The causes mentioned by Wex of the hydraulic changes under discussion were partly combated and partly doubted in several respects, and particularly the assertion that in cultivated countries the rainfall has diminished by reason of the clearings; but it was admitted that the clearings had exercised a very injurious influence on the regimen of rivers.

The Society of Austrian Engineers and Architects adopted the above resolutions of its committee of experts at its business meeting on April 17, 1875, and authorized its executive committee to submit the amended conclusions of the Court Counselor, Wex, to the Imperial Royal Austrian Government, with the request for their execution, in order to limit the threatening calamity as much as possible.

6. At the International Congress of Agriculturists and Foresters at Vienna, in September, 1873, the royal Prussian inspector of forests, Dr. A. Bernhardt, related, and then the royal Italian senator, Louis Torelli, of Rome, submitted, very interesting data concerning the results of clearings in Europe, and proved thereby that many springs fail; that the discharge of creeks and rivers at a normal stage decreases more and more; and that on the other hand floods now occur oftener and rise to a greater height.

7. Mr. Torelli, particularly in his work* published May 10, 1873, furnishes data worthy of notice, gathered from the copious hydraulic observations and notes taken in Italy, showing the injurious effects of clearings in that country, and gives the following opinion, based upon abundant examples and experiences, viz:

"However great may be the evils and disadvantages of the more frequently and higher rising floods of the present time, they are not greater than those which result from the progressive failing of springs and decrease of volume of water in creeks and rivers."

* "Delle cause principali delle piene dei Fiumi e di alcuni provvedimenti," per diminuirle di Luigi Torelli, Senator del Regno. (The principal causes of high waters and some measures for diminishing them, by Louis Torelli, senator of the kingdom.)

Mr. Torelli further cites the following observations and measurements of the distinguished expert, Paleocapa, viz:

The volume of water at the lowest stage of the river Sele has decreased 33 per cent. during the last 150 years; that of the river Brenta, at Bassano, 7 per cent., between 1684 and 1877; and finally that of the river Adda, where it flows out of Lake Como, 13 per cent., between 1842 and 1862; and this decrease still continues, and therefore creates apprehension.

8. In consequence of the destructive inundation of the river Po, in 1872, the Royal Italian Government organized a commission consisting of seven experts, whose duty it was to make a minute hydraulic survey of its whole valley, and, after a thorough study thereof, to recommend such measures as would be suitable to prevent similar occurrences.

In the report made in December, 1876, by Mr. Basilari, vice-president of the supreme board of public works, the commission expressed the following opinion, viz:

The floods of the river Po have constantly increased in height, especially during the present century, in consequence of which it became necessary to raise and strengthen the greater portion of its levees. This method of relief was considered by the commission not only as inadequate, but as really increasing the danger, for which reason it was led to consider means by which the height of these floods could be diminished, or, at least, the increase stopped. Consequently it examined into the condition of the forests in the valley of the Po. It laid great weight on the enactment of suitable laws against destruction of forests; recommended next the construction of large reservoirs for the storage of a portion of the floods, which might subsequently be used for irrigation; further proposed the construction of suitable works for improving the mouths of its tributaries; and finally recommended several cut-offs, in order to diminish the length of the stream.

The preceding opinion and the proposed measures therefore fully coincide with the views expressed in my treatise of April, 1873.

9. Professor Dr. Ebermayer, in his admirable work entitled "Physikalische Einwirkungen des Waldes auf Luft und Boden und seine klimatologische und hygienische Bedeutung begründet durch die Beobachtungen der forstlichmeteorologischen stationen in Königreich, Bayern. Aschaffenburg, 1873,"[*] has furnished convincing proof of the following statements, based upon minute scientific experiments and observations during many years, viz:

(a) The forest influences the amount of rainfall by increasing the relative humidity of the atmosphere and bringing it nearer the point of complete saturation, so that when the temperature is lowered a partial separation of the water takes place more easily and in greater quantity than in clearings. This influence is more marked in proportion to the height of the forest above the level of the sea.

(b) The forest diminishes the evaporation of the surface water in a much greater degree than an open field. In those whose grounds are kept clean it amounts to 62 per cent., and in the others to 68 per cent.

(c) In the interior of the continent, where the humidity of the atmosphere and annual rainfall decrease and the degrees of summer heat increase, the forest has a greater influence on the rainfall than in the coast regions. Ireland and Great Britain can dispense with their forests more easily, as far as rainfall is concerned, than Germany or Russia.

[*] The physical effect of the forest on the atmosphere and soil, and its climatological and hygienic importance, based upon observations made at the foresters' meteorological stations in the Kingdom of Bavaria. Aschaffenburg, 1873.

(*d*) Large clearings in level countries have material influence, it is true, but in hilly regions there will be, on an average, less rainfall than before, and especially in the warmer half of the year.

Even if it were assumed that, after clearing forests, under all circumstances, the rainfall would be the same as before, the enormous influence of the forest, and its ground covered with leaves and brush upon the evaporation of its surface water, in consequence of which, compared to the surrounding country, it must be considered a great reservoir, would sufficiently explain the well-known failing of springs and its influence in diminishing the medium stages of the rivers.

(*e*) There is another fixed relation between the stage of water in rivers and the forest caused by the fact that in wooded localities the snow lies three or four weeks longer and melts less rapidly than in clearings. In an open country the rise in rivers occurs more rapidly after the melting of the snow; the latter runs off more rapidly, less soaks into the ground, and the springs are therefore fed less than in wooded countries.

(*f*) Extensive clearings are more injurious in hilly than in level countries. They produce in a short time destructive torrents, dry periods, then short but destructive inundations, formation of sand-bars in rivers, want of perennial springs and creeks, and great variations in the stage of the rivers. These most certainly occur and characterize such regions in which the hills have been cleared.

Mr. Ebermayer says, finally, viz:

(*g*) "If we combine the different effects of the forests, there can be no doubt but that their disappearance must considerably diminish the volume of water in a country, even if after the clearing there is as much rainfall as before."

"The preceding statements show how closely the wealth of forests and water in a country are bound together; a circumstance which is particularly due to the powerful influence which the forest and its ground covered with leaves and brush has upon the evaporation of the moisture in the ground. It is, then, not to be wondered at that springs and creeks dry up or flow only periodically; that the mean stage of rivers and creeks diminishes in height when large clearings are made in a country, and that on the other hand springs flow more copiously and regularly when new trees are planted and forests are increased in extent."

10 and 11. The imperial chief forester, Edward Ney, of La Brogue, in Alsace, in a paper published in January, 1875, entitled "Ueber den Einfluss des Waldes auf die Bewohnbarkeit der Länder,"* and Dr. Jacob von Bebberer, in his paper published in 1877, entitled "Die Regenverhältnisse Deutschlands,"† both written after observations made during a long series of years, on the great influence of the permanence of forests on the hydraulic condition of countries, have expressed precisely the same opinions as Professor Ebermayer.

As an illustration of the fact that the clearing of hilly countries frequently results in the complete failing of springs, Mr. Ney mentions that in the Provence, after all the olive-trees, which there formed regular forests, had been frozen in 1822 and cut down, a great number of springs failed totally, and that besides, in the city of Orleans, after the surrounding heights had been thus cleared, nearly all the wells dried up, and it became necessary to conduct the headwaters of the river Little Loire into the city.

Mr. Ney furnishes the following examples to prove that during the

* "Concerning the influence of the forest on the habitableness of countries."
† "The circumstances relating to rain in Germany."

historical period many rivers once rich in water supply have become much poorer therein, notwithstanding their more frequent floods in consequence of the destruction of the forests, viz:

At the time of the Roman rule in France, the river Durance, south of Avignon, and the Seine were navigable rivers and richly supplied with water, so much so that the navigators of the Durance formed an influential corporation, and Emperor Julian, who resided in Paris during a period of six years, particularly extols the constant even stage of the Seine. At present, since the regions of the headwaters of these two rivers have been cleared, the Durance can hardly float a skiff in summer, and the Seine, in which the difference between the high and low water stage is now 32 feet 10 inches, was only made navigable again by the construction of numerous wing-dams.

12. The director of the royal board of foresters of Hanover, Dr. Burckhardt, in a letter addressed to me on March 5, 1876, says the views expressed by me in my treatise were expressed in words which were as if taken from his own soul, and that he had also, in the course of his observations extending through a period of twenty-five years, seen the sad results confirmed that in consequence of the clearings made in North Germany there resulted the disappearance of many lakes and ponds, draining of marshes, failing of many springs and creeks, and the lowering of the surface water, whereby the cultivation and productiveness of the remaining fields suffered sadly.

13. Prof. Alexander Bettochi, inspector of the Royal Corps of Civil Engineers at Rome, in his memoir submitted to the Royal Academy, "dei Lincei," in June, 1876, which was subsequently published as a paper, acknowledges that my theory relative to the diminution of water in springs and rivers contemporaneously with the increase in height of floods agreed with observed results, and still further substantiated it by plotted observations made on the Theiss at Szegedin.

14. Mr. Frederick Symong, imperial and royal professor of geography at the University of Vienna, honorably known by his researches and surveys in the Alpine regions, in his lecture entitled "Schutz dem Walde,"[*] given February 21, 1877, before the Society for the Diffusion of Knowledge of Natural Sciences, at Vienna, acknowledged my theory as being correct, and at the same time proved conclusively that in consequence of the destruction of the forests in mountainous regions the rainfall diminishes at those points, vegetation on the mountain sides is constantly receding, the soil is washed off, after which the mountains gradually become perfectly barren. A further result of this is a decrease of water in the lower strata, in springs, creeks, and rivers, which in turn are succeeded by the drying up of the valleys, and the gradual devastation of cultivated fields, thus proving the truth of the saying of a great naturalist, "Man strides over the earth and a desert follows him."

15. Dr. F. W. Dunkelberg, director of the Royal Agricultural Academy at Popelsdorf, in his work entitled "Die Schifffarths-Canäle in ihrer Bedeutung für die Landes Melioration"[†] (Bonn, 1877), concurs with me in my theory relative to the decrease of water in springs and rivers, it being the result of his own observations and experience.

16 and 17. Further along during my discussion of the observed results on the Elbe, I will give the concurring views of Mr. M. W. Schmidt, of Dresden, royal director of hydraulic works, and of Mr. Maass, of Magdeburg, a member of the royal board of public works.

---

[*] Protect the forest.
[†] The importance of navigable channels in the improvement of countries.

18. Mr. Robert Lauterburg, the Swiss engineer and hydraulic expert, who occupies a great deal of his time in collecting exact data and in comparing the relative discharge of the rivers of Switzerland, in his paper entitled "Ueber den Einfluss der Wälder auf die Quellen und Stromverhältnisse der Schweiz"* (published by Schulze at Basle, 1877), announces the following very reliable results of his surveys, viz:

In the Melasse formation within an area of .29 of a square mile the relative discharge of springs was accurately measured, and it was found that those in wooded portions discharged from five to ten times the amount of those in the clearings.

Now, as it is universally known that during those periods in which for weeks there is no rain the creeks and rivers are almost exclusively fed from the springs, filtration, and surface water, and also known that since the introduction of railroads and telegraphs in Europe, as well as the general and extraordinary increase in all kinds of manufactures and trade, such a colossal consumption of building and fire wood has taken place that in Europe several million hectares† of forest which formerly existed, and particularly in hilly regions, have been cut down and destroyed, it can be concluded in advance, from the result of these observations in Switzerland, and without giving the gauge readings on rivers and streams the least consideration, that during the last forty years the discharge of springs, creeks, and rivers, during low and medium stages, must certainly have decreased.

Now, I will take the liberty of showing that not only in Europe, but in the cultivated portions of other parts of the world, the discharge of their rivers and streams has also decreased during the last decade.

19. In a report made to the House of Representatives on March 14, 1874, the Commissioner of Public Lands of the United States of America expressed his fear that the continued destruction of the forests would seriously injure the public interests. At the request of the Commissioner to submit this important question to thorough experts, the American Academy of Sciences charged a committee composed of several of its members with the task. After the committee had collected accurate data and given the subject mature deliberation it made an extensive report, full of statistical information, from which I will quote only the following chief points.

The committee in this report recounts the bearing and importance of forest culture not only for timber, but also as a means of improving the general welfare, as the climatic conditions depend upon the existence of forests and deteriorate with the destruction of the latter.

The evident results of the destruction of forests are "the failing of springs, drying up of creeks, decrease in volume of water in the rivers, canals, and streams, and the increasing difference between high and low water stages in the latter." The fact that the volume of water in rivers and streams decreases in proportion to the destruction of the forests will not escape any careful observer, and America is threatened with the danger that changes will take place in the permanence of its largest rivers even, fraught with the most serious results, unless suitable measures are adopted in proper time.

The committee finally enumerated and recommended measures and precautions most suitable to stop this threatening calamity, which are nearly the same as those which I proposed in my first treatise of 1873.

Congress, appreciating these suggestions, passed an act dated August

---

* On the influence of forests on the conditions of the springs and streams of Switzerland.

† About two and one-half acres.

15,1876, appropriating $60,000 for the purchase of seeds and plants and generally for the improvement and promotion of forest culture; and a further sum of $2,000 as a prize to an expert in the investigation of the question of forest and tree culture and the means which best commend themselves for the preservation, improvement, and planting of forests in North America.*

20. The above prize was won by Dr. Franklin B. Hough, through his scientific work† on this subject, in which he points out in great detail the influence of forests on the even and seasonable rainfall, upon the acquisition of a regular water supply to the springs, creeks, and rivers, and finally on the prevention of extraordinary floods, and in which he then cites the proofs furnished by my treatise of April, 1873, and agrees with me in the views therein expressed and with the theory concerning the decrease of water in springs and rivers as therein announced.

It will be seen from the official proceedings and the scientific data which have been collected, as just mentioned, that already in North America even, where, partly through the vandalism, ignorance, or avarice of the inhabitants and partly through elemental disturbances, very extensive tracts of wooded lands have been cleared and totally destroyed, the highly injurious effect thereof, and particularly the decrease in the volume of water in their springs and rivers, probably has become more evident than to us in Europe, and that also the government of that country has forthwith recognized the danger resulting from this not only to its present but also to its future generations, and that it now proposes to apply in the most energetic manner such measures as will prevent this danger as much as possible.

21. The only knowledge which I have thus far obtained relative to to this question in South America is taken from the following article, dated December 5, 1878, written from Rio Janeiro to the Cölnische Zeitung,‡ viz:

"A phenomenon is developing which causes apprehension on the part of the inhabitants on the banks of the Amazon, namely: The stream is receding in an appalling manner, and particularly above Manaos navigation is already impossible. There is a constant decrease in the stage of the river, the causes of which are wholly unknown. It is very desirable that thorough naturalists should examine this phenomenon."

Now, if such a remarkable decrease in the volume of water has been observed in the upper portion of the Amazon, which is the largest and has a greater wealth of water than any other river on the face of the globe, it is very probable that similar decrease in the volume of water has taken place in the other rivers and streams of South America, and that this cannot be substantiated for want of observations alone.

22. Mr. John Croumbie Brown, the distinguished naturalist, formerly professor of botany at Cape Town, in his book entitled "Hydrology of South Africa, compiled by John Brown, LL.D., Kirkcaldy, printed by John Crawford, 201 High street, 1875," has published very interesting

*The author is mistaken herein. The act approved August 15, 1876, directed the Commissioner of Agriculture to appoint some man of approved attainments and practically well acquainted with the methods of statistical inquiry, with a view of ascertaining the annual amount of consumption, importation, and exportation of timber and other forest products, the probable supply for future wants, &c. On August 30 1876, Hon. Frederick Watts, then Commissioner of Agriculture, appointed Dr. Frank, lin B. Hough, of Lowville, Lewis County, New York, to the discharge of this important duty.

†Report upon forestry, prepared under the direction of the Commissioner of Agriculture, in pursuance of an act of Congress approved 15th of August, 1876, by Franklin B. Hough. Washington, Government Printing Office, 1878.

‡Cologne Gazette.

observations and experiences, extending through a long series of years, concerning the former and present condition of the countries as well in Africa as in the larger islands of the Southern Ocean, of which I will briefly mention but a few.

In South Africa there are yet found scattering groups of trees of gigantic dimensions and great age, but without any corresponding after-growth, which are undoubtedly the remnants of former extensive forests which were cleared or destroyed by fire. These ancient gigantic trees prove the former great humidity of the climate and the rank fertility of the soil.

The destruction of the forests in South Africa and the custom which prevails there of destroying their tall grasses by fire have contributed very much to the parching of the soil in those localities, so that new trees are only found along the banks of rivers and in high mountain passes.

Dr. Livingston and Dr. Moffat also describe the burnings just mentioned, and the latter refers to an extensive forest of wild olive trees in the vicinity of the city of Griqua which was destroyed by fire, and mentions the gradual decrease of rainfall which resulted therefrom.

Mr. Brown notes many observations concerning the numerous and sudden changes which occur at present from severe droughts to violent rains, causing destructive overflows and the evident decrease in the volume of water, indeed the complete drying up of many creeks and rivers as well as their change into torrents; which, however, are only created when the forests and plants of the mountain slopes were destroyed and the soil was washed away in consequence thereof. In countries where mountains are covered with forests no torrents are formed, but, on the contrary, plenty of springs, creeks, and rivers, which serve to increase the fertility of the soil.

The author further reports upon the decomposition of the former rich soil by the sun's rays, by which the country becomes sterile, and the fields which were once rank are converted into deserts, and says that these very disastrous changes are, according to the scientific investigation of Mr. Brown and a naturalist cited, by him, entirely due to the destruction of the forest in those countries.

In the concluding chapter of his work Mr. Brown recommends, for the purpose of removing or at least ameliorating the evils which have resulted from the destruction of forests, measures and precautions almost identical with those proposed by me in my first treatise of 1873, although the latter was unknown to him when he compiled his work.

In the second work published by Mr. Brown, entitled "Forests and Moisture, or Effects of Forests on Humidity of Climate, compiled by John Croumbie Brown, LL.D., Edinburgh, 1877," he proves conclusively, in more detail, not only theoretically but also by numerous illustrations, the fact that forests have a great influence and effect on the humidity of the air and surface of the earth, the drying up of marshes, the formation of clouds, on the quantity and regular distribution of rain among the different seasons, and the discharge of springs and rivers. In this work Mr. Brown cited the data collected in my treatise of 1873, and declares himself as agreeing with me in my theory concerning the decrease of the volume of water in springs and rivers.

23. In conclusion, I must here advert to a fearful catastrophe which lately occurred, and strikingly proves the disastrous results of unlimited destruction of forests.

The northern province of the Chinese Empire, Shan-Si, whose capital is Tayenn, is inclosed on all sides by high mountain ranges, which in early ages were covered with dense forests. At that time it rained here every year periodically, the atmosphere was sufficiently humid, and it belonged to the fertile, well-cultivated, and densely-settled provinces of the Chinese Empire. But the inhabitants of this once blooming and happy country, through greed and in the endeavor to increase the yield of these mountain slopes more and more, have gradually cleared the surrounding mountains completely. The result is that the former periodical rains have almost disappeared, and that the rainfall as well as the humidity of the atmosphere have decreased, and that, consequently, in this province failure of crops, want, and misery have succeeded each other during several years. In 1877, already, there was such a general drought, failure of crops, and famine, that in consequence thereof nearly three millions of souls perished. In the official report of the governor, Li Ho-nien, chief commissioner of the board of relief, the following account of this catastrophe appears, viz:

"During the first period of this unheard-of famine, the living subsisted upon the corpses of the dead; then, later, the weak were devoured by the strong; now the misery has reached such a stage that the people devour their own blood relations. History, up to this date, has not shown a more horrible state of affairs than this, and unless measures for relief are promptly taken the entire population of this section of the country will be destroyed."

The Catholic bishop, Monagatta, in a letter from Tayeun, dated March 24, 1878, corroborates the description contained in the above report of the horrible scenes enacted during the famine.

The ruinous results of the destruction of forests are also showing themselves in the other provinces of China; in the highlands by killing droughts, and simultaneously in the southern lowlands of the empire by destructive rains and overflows.

CHAPTER II.

CLEARING UP THE DOUBTS AND REBUTTING THE ARGUMENTS OF THE OPPONENTS TO MY THEORY ON THE DECREASE OF WATER IN SPRINGS AND RIVERS, AND COMMUNICATING THE NEW DATA AND OBSERVATIONS RELATING THERETO COLLECTED BY ME SINCE 1873.

So far as I have been able to learn, up to the present time, from published works and scientific periodicals, the following engineers and experts have expressed views in opposition to my theory, viz:

1. The committee of experts of the Society of Austrian Engineers and Architects, which, in its report already referred to by me, agrees with my theory in general, but doubts somewhat the proofs which I offered and the conclusions drawn therefrom.

2. Mr. Sasse, royal Prussian ministerial counsellor and member of the board of public works (Zeitschrift für Bauwesen, von G. Erbkam, in Berlin, von Jahr 1874). *

3. Mr. Kluge, royal Prussian inspector of hydraulic works (in the journal just mentioned).

4. Mr. Schlichting, royal Prussian inspector of hydraulic works (Deutsche Bauzeitung, Jahr 1876). †

5. Mr. Grebenau, member of the imperial board of achitecture and director of hydraulic works (Deutsche Bauzeitung, Jahr 1876.) †

---

* Journal of Architecture, Berlin, 1874, G. Erbkam, publisher.
† German Gazette of Architecture, 1876.

6. Mr. Grave, royal Prussian district architect and deputy director of public works (Deutsche Bauzeitung, Jahr 1877.) †

7. Mr. Charles Hurich, royal Hungarian ministerial counsellor (Zeitschrift des ungarischen Ingenieur und Architecten Vereines.) *

8. Dr. Joseph Ritter Lorenz von Liburnan, imperial and royal ministerial counsellor (Wald, Klima, und Wasser, 1878.) †

These opponents urged against my theory and the proofs on which it rests the following doubts and arguments in general, viz:

A. That the decrease in the stages of water, i. e., the lowering of the water surfaces shown by me to have taken place at the nine gauges on the five rivers may not have resulted from a decrease in discharge, but from a deepening of the river beds or a change in the slope caused by works of improvement.

B. That the established fact that the stages of these rivers have decreased in height does not furnish reliable proof that their discharge has decreased.

C. That it may be possible that the increased discharge of streams, produced by the higher floods of the present time, brings the decreased discharge caused by the diminished height of low and medium stage to the former general average, and then the observations furnished would only show that during the last decade the regimen of rivers and streams has undergone a change.

D. That gauge readings of a stream extending through a period of at least 200 years are necessary in order to draw a reliable conclusion from them whether a decrease in volume of water has really taken place or not.

E. That it cannot be determined with certainty from a comparison of the heights of different stages of streams whether a decrease in the volume of water has taken place or not, and that this can only be done by directly measuring, from time to time through a long period of years, the actual discharge of rivers and streams.

F. That the assertion made by me that the amount of rainfall had decreased in consequence of the great devastation and destruction of forests is not at all proven by the examples which I furnish, and the opinions of scientific authorities which I cite, because from meteorological observations extending through a long series of years in England, at Paris, St. Petersburg, and Copenhagen it cannot be deduced that the rainfall has diminished.

G. Mr. Grebenau declares that my theory relating to the decrease of volume in water is incorrect, and advances instead the new theory that the beds of creeks, rivers, and streams, are continually deepened by the erosive power of the running water, and that alone in consequence of this deepening, the heights of the water stages constantly decrease, i. e., the water surface of rivers and streams sinks deeper and deeper.

Since Mr. Grebenau declares my theory incorrect, and substitutes therefor one which is entirely new in hydraulics, I propose, first, to thoroughly discuss the arguments and proofs upon which it is based, because, in case his assertions and theory should be well founded, the discussion of the other doubts and objections can be dispensed with.

In consequence of the invitation heretofore mentioned, given by the Imperial Academy of Sciences of Vienna, to all foreign scientific institutions, Mr. Grebenau, among others, was requested to give his opinion of my treatise on the decrease of water in rivers and streams, based upon his observations and experience relating to this subject.

---

* Journal of the Society of Hungarian Engineers and Architects.
† Forest, climate, and water.

Upon this, Grebenau, with his characteristic tireless diligence and energy, collected the gauge readings of fourteen larger rivers and streams, compared them, combined them in various ways, at the same time studied zealously the water question raised by me, and then gave a detailed lecture as the results of his investigations, on September 6, 1876, at the general meeting of the Union of Societies of German Engineers and Architects, held at Munich, on the "Sinking of rivers, and the phenomena accompanying it," in which he communicated the following, viz*: He compared the readings of 75 gauges on 14 rivers for low, medium, and high water stages, divided them into two periods of observations, similar to the manner in which Mr. Wex did, calculated the mean of these, compared them, and, assuming that the mean of the observations at the above-mentioned stages need always only be taken into consideration, obtained the following results, viz:

The medium stage of 12 rivers, supplied with 67 gauges, receded, i. e., the surface of the water sank, on an average, from 3 feet 3¾ inches, to 6 feet 6¾ inches; on the contrary, the stages of 4 rivers at 6 gauges increased in height, and, indeed, just at those at which the half periods of observation were very short, they lasting only 7, 7½, 12½, and 16½ years.

From the results of these observations Grebenau concluded that in the changes which take place in the fall and rise of the river at several gauges on the same river at the same time of observation when the discharge is nearly uniform, no decrease in the volume of water can be substantiated, because if the fall of the river shown by the reading of one gauge would indicate a decrease, a rise shown at another gauge of the same river would indicate an increase in the volume of water, which clearly is impossible.

"Now, as Mr. Wex admits in his treatise that on several rivers, and particularly at nine gauge stations on the Danube, the bed of the river either raised or sunk, his theory that a decrease in the height of stages of a river or a sinking of its surface indicates a decrease in the volume of the water becomes, through this admission alone, untenable."

Grebenau further remarks concerning my theory that the great amount of the decrease in the stages of several streams established by the figures given in my treatise prove that this cannot be the result of a decrease in the volume of water, because otherwise these streams would in several hundred years be without water and their beds dry.

In reply to these conclusions of Grebenau, I believe it to be my duty to make the following explanation:

It is certainly generally known that the beds of rivers and streams which are not yet improved, or on which the improvements have just been begun, are deepened in some stretches and raised in others by the formation of bars, and that therefore in consequence of this a sinking or raising of the water may occur, from which no reliable conclusions could be drawn.

But if, from the numerous gauge readings which are submitted by me are eliminated those which were taken on stretches of the stream in which changes in the bed of the river took place, we will still find some rivers or stretches of streams which lie either in a natural unchangeable

* An abstract of the letter was published in the German Gazette of Architecture, of October 21, 1876. Grebenau had more extensively elaborated his investigations in the water question, and had prepared them for publication when, unfortunately, too soon death tore him away on June 23, 1877. His highly-respected widow was so kind as to transmit to me the manuscript for inspection; but our united efforts to publish it were vain, in consequence of the necessary expense, which was considerable.

bed or which have been improved from time immemorial and are in a permanent condition. The most scrupulous expert must admit that on such rivers and stretches we can justly assume that the decrease in their stages, *i. e.*, sinking of their surface, indicates a decrease in the volume of water, since it would be impossible to explain the phenomenon in any other way.

I must remark concerning the second objection of Grebenau to my theory that in my first treatise of 1873 I only expressed the apprehension lest the creeks and rivers, still replete with water, which exist in cultivated countries might, gradually, during a larger part of the year have little discharge; but, on the contrary, after heavy rains swell fearfully and become converted into torrents.

It is a historically-established fact that such changes have already taken place in the cultivated countries of ancient times, and only those can doubt it who are unacquainted with the geography and history of those countries.

It is out of the question to suppose that a stream which is created by the confluence of many creeks and rivers can ever completely dry up, because, although the rainfall is diminished by clearings, yet it will never completely have an end, and because in a large river valley the springs, creeks, and tributary rivers can never lose their water simultaneously, since the rainfall is irregularly divided among the valleys of the different creeks and tributary rivers, and the high water of the one frequently reaches the main stream simultaneously with the low water of another. The change of the tributaries into torrents cannot therefore completely dry up the main stream, but it may cause great and numerous variations in its stage, which might become excessive. It is clearly apparent from the plot of gauge readings given in my first treatise that this change has already begun in several streams of Europe.

I believe that I have thus successfully controverted the objections brought forward by Grebenau against my theory, and I will now pass to a closer examination of the new theory established by him.

Grebenau, as already stated, pointed out a considerable decrease in the lower, medium, and higher stages, *i. e.*, a sinking of the water surface of 12 rivers at 67 gauges, and explained this remarkable phenomenon by the following new theory, viz:

The mountains on the surface of the earth are being continually reduced in size by washings caused by the working power (called by later geologists the power of erosion) of running water and the adjacent depressions or valleys filled up and elevated. Grebenau then sets up the following thesis on the effect of this power of erosion of running water on the beds of streams:

1. The mud which is created when a river rises is due to the friction of the different kinds of *débris* upon each other, and the floating mud is therefore the necessary result of the rolling *débris*.

2. This mud is generated and carried farther at each point when there is rolling *débris*, and, therefore, increases continually in a downstream direction; and, indeed, according to the rule of arithmetical progression.

3. The voids produced in the *débris* by the mud thus carried off are the causes of the continued and lasting deepening of the river bed and the resulting sinking of the water surface.

Grebenau, assuming these theses as established in discussing the Rhine, whose water surface in Alsace, as he proves, sinks on an average .64 of an inch annually, and, therefore has sunk 5 feet 4 inches during

100 years, and 53 feet 4 inches during the last 1,000 years, arrives at the conclusion that about 5,900 years ago the falls at Schaffhausen did not exist, and have only been caused since that time by the sinking of the bed of the stream below, and that about 1,000 years ago the surface of its central portion below Basle was at the height of its present high banks, or 53 feet 4 inches higher than it is, and that it then washed and formed these banks and since then has deepened its bed and lowered its water surface to that amount.

If the theory of Grebenau that the beds of all rivers and streams are deepened by the power of erosion at a continuous rate of ⅛ to ¾ of an inch annually—consequently from 3 feet 1½ inches to 6 feet 3 inches in 100 years, and from 31 feet 3 inches to 62 feet 6 inches in 1,000—were considered well founded and true, it would have a very injurious effect upon the cultivation of countries and future generations. I have, therefore, since 1876, collected data and studied the question in the most thorough manner, and I take the liberty to communicate briefly the result thereof.

It is known to every hydraulic engineer and geologist that rivers and streams flowing through large broad valleys, if left to themselves, do not d. epen their beds, which consist of coarse rounded gravel and *débris*, but cause their banks, which consist of lighter earthy material, to cave in, largely increase the width of their beds, create islands, bends, and swift currents, and finally, as they lose through this the most of their transporting power, the *débris* from the regions of their headwaters and tributary creeks and rivers is deposited and raise their beds.

During past ages these wild rivers and streams meandered at will in the valleys, and created their existing banks by filling up the deep gulleys and basins of the sea which previously existed. At the present time these wild streams raise their banks by deposits of mud during overflows, but not as rapidly as their beds are raised, and it frequently happens that the difference of level between the banks and bed of the river is diminished and the height of overflows increased.

It is also very well known to every experienced hydraulic engineer that the velocity and motive power of these wild streams can be so much increased by their improvement, i. e., by narrowing its wide stretches, closing subsidiary channels, making cut-offs, and finally protecting its banks, thus increasing their slope; that the beds of these improved portions will be gradually deepened by the removal of the pebbles and *débris* which raised them, and that they will be enabled to move the *débris* which is brought from the region of the headwaters by high waters farther down stream. If, however, this improvement is not made throughout the whole stream this *débris* will be deposited in the lower parts thereof, and thus raise its bed and cause it again to run wild.

The fact just mentioned that rivers and streams, in their natural state, have a tendency to raise their beds, and consequently their water surface, and that a deepening of the former and lowering of the latter can only result from their thorough improvement, has been already mentioned in a detailed manner by Mr. Hagen, royal Prussian privy counselor of the superior board of public works, in his admirable manual on the knowledge of hydraulic works, and many more of the distinguished hydraulic engineers of Germany and Italy who I requested in writing to give me the result of their observations on this point have declared that after many years devoted to the study thereof they consider this an established fact, and none of them could furnish me with a single instance in which the bed of a river or stream was continually deepened by the natural power of erosion of the running water.

Although it must be admitted that the friction of the particles of gravel brought down by high water against each other does cause some mud, yet it must be clear that by far the greatest part of that carried along by high water is produced by the rain washings of the valleys, by caving of unprotected banks, and finally by the *débris*, sand, and earthy matter emptied into the stream by its tributary creeks and rivers, so that a deepening of the bed cannot be even inferred from the quantity of sand carried off by a stream, much less can it be measured thereby, because that which is produced by the friction among the particles of *débris* is vastly overweighed by the very large quantity of rubble sand and earth which is brought in at every high stage from the headwaters and tributaries.

From this it follows that the theory advanced by Grebenau that the mud transported by the high water of streams is principally created by the friction of the particles of *débris* on each other, and its sequence that the beds of old rivers and streams are continually deepening through the power of erosion of their running water, is evidently incorrect, since in all rivers and streams in their natural state precisely the opposite, namely, the raising of the bed, takes place.

The incorrectness of the theory advanced by Grebenau is also proven by historical facts, for if the deepening of river beds had continually taken place in earlier times all valleys whose surfaces are now only from 6 feet 6¾ inches to 9 feet 10½ inches above the medium stage of the river must have been permanently inundated about 500 years ago, which plainly was not the case, because at that time cities and many inhabitated places existed therein. Mr. Grave in his criticism of Grebenau's theory* (Deutsche Bauzeitung von Jahr, 1877, Nos. 54 and 56) has already proven at considerable length the fact that it was opposed to historical facts.

I can leave the contradiction of Grebenau's theory concerning the creation of the falls of the Rhine by reason of the sinking of the river bed below to the geologists, who are better versed in such a subject, but I must, on the other hand, opppose in a determined manner is further assertion that about 1,000 years ago the water surface of the Rhine below Basle was on the same level with its high banks, and that it made these, and since that time its bed has been deepened about 53 feet 4 inches.

Grebenau, at the general meeting of the "Pollichia," in Dürkheim, delivered a lecture, on September 11, 1869, entitled "Der Rhein vor und nach seiner Regulirung,"† which was subsequently published as a paper, with maps and profiles attached, which proves, at great length and very thoroughly, that in prehistoric times there existed, in the present valley of the Rhine, between Basle and Bingen, a large sea which gradually flowed off, as the mountain gorge between Bingen and Bonn was formed. The basin of this sea was filled up by the large masses of rubble and *débris* brought into it, not by the Rhine alone, but also by the creeks and other rivers which emptied into it, to within from 32 feet 9¾ inches to 65 feet 7½ inches of the present surface of the valley. The high banks, also called the Dilurian terraces, which are from 3.1 to 4.3 miles apart and 32 feet 9 inches in height, were clearly once the rim of this basin, and received their present form partly through the action of the waves of this sea, which subsided very slowly and in proportion to the crown

---

* German Gazette of Architecture, 1877, Nos. 54 and 56.
† The Rhine before an l after its improvement.

of the rocky crest of the falls at Bingen, and partly to the undermining caused by the Rhine, which meandered in this irregularly filled-up basin in the wildest manner. Since the basin has filled up, and after the slope had become regular between the crest of the overfall at Bingen and the river bed, also consisting of rock, at Basle and at Waldshut, the Rhine has not only not lowered its adjacent banks, which are composed of coarse gravel, but on the contrary has raised them from 3 feet 3¾ inches to 6 feet 6¾ inches, by depositing upon them sand and mud during its frequent overflows.

The gradual sinking of the water surface of this sea is, to-day, plainly visible on the high banks on both sides, and also the subsequent undermining of the latter by the wildly meandering Rhine, and then, in consequence of its improvement, begun in 1817, between Basle and Manheim, by straightening and shortening its length from about 165 miles to 111 miles, a deepening of the obstructed river bed and sinking of the water surface of from 3 feet 3¾ inches to 4 feet 11₁⁄₆ inches took place in some stretches of the river. The excellent hydraulic engineer, Grebenau, observing these visible facts, was led to the erroneous conclusion that this deepening of the river bed and sinking of the water surface were the result alone of the power of erosion of the Rhine, and, as he at the same time found that at 67 stations on 12 rivers, during a long period, there had been observed a considerable decrease in height of the low, medium, and high-water stages, Grebenau advanced the incorrect theory that, since the existence of all creeks, rivers, and streams, without exception, their beds had been continually deepened, and their water surfaces sunk by the power of erosion of the running water, and that this erosion will continue steadily in the future.

In saying the preceding I do not wish to cast any reflection upon the memory of my departed, highly valued, and personal friend Grebenau, who was known and respected in the whole of Germany as a distinguished hydraulic engineer, but I simply desire to clear up a very important hydraulic question.

Now I will attempt to clear up, i. e., to rebut, the arguments, objections, and doubts heretofore enumerated of my other opponents, and for that purpose will refer to the data and observations collected by me since 1873, and which strengthen the correctness of my theory and prove Grebenau's utterly untenable.

Reply to A.—If the decrease in height of the stages of a stream, i. e., the sinking of its water surface which was observed during a long period, were alone the result of the deepening of its bed, either by the power of erosion of the running water or by the executed improvement of the stream, these deepenings must take place in a large part of the stream, must be created successively and in equal degree, and finally must be permanent.

The changes in the bottom of the bed of a river, such as partial washing out by whirlpools and shifting of the channel, which occur at almost every high stage cannot be considered as being a deepening of the river bed in the sense in which it has been herein discussed, as these are generally corrected again by the next high water, and consequently can have no effect upon the monthly or yearly average of the guage readings.

In order to judge clearly and in a reliable manner of the effect which the deepening of the bed of a stream through a long stretch has upon the water stages thereof, it is necessary to plot the original cross-section of this stream, together with the deepening in the bed and various stages thereof, and we shall then arrive at the following unassailable conclusions:

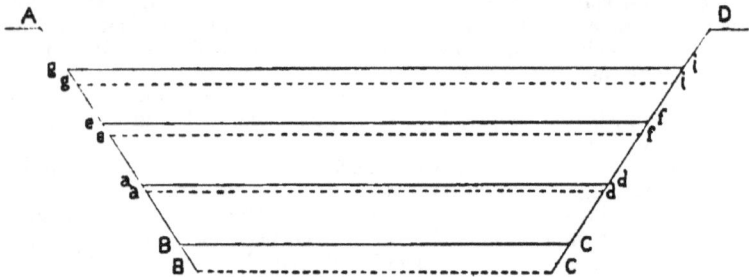

In the above figure let A, B, C, D represent the whole cross-section and profile of discharge of a stream, which for the sake of simplicity we can assume as perfectly regular and with sloping banks. In this section let the line $a\,d$, $e\,f$, and $g\,i$ represent respectively the low medium and high stages of water.

If, then, after a lapse of a long period of time, say from twenty to thirty years, the bottom of the original bed B C has been deepened to the new bottom B C, either in consequence of the power of erosion or the improvement of the stream, we can calculate the resulting amount which the surface of the water has sunk, provided we assume that the volume of water has remained the same.

Denoting the width B C of the bottom of the bed of the stream by b, the velocity of discharge by r, and the distance which the bed has sunk by h, the amount of discharge, m, through the deepened portion of the bed B B, C C, we will have

$$m = b\,h\,r.$$

Now, denoting the width of the water surface at low, medium, and high stages respectively by b, b_2, and b_3, the mean velocity at the surface at these stages respectively by r, r_2, and r_3, and the sinking produced by the deepening of the bed respectively by h, h_2, and h_3, it follows that since the volume of water is only reduced by the amount which flows through the deepened portion B B, C C, the following equations are true, viz:

$$m = b\,h\,r = b\,h\,r = b_2\,h_2\,r_2 = b_3\,h_3\,r_3.$$

Now, since in the cross-section of all streams, excepting in the rare case where both banks are reveted with vertical walls, the width of the water surface increases with the height of the stage, that is, $b_2 > b$ and $b_3 > b_2$, and since, as is well known to every hydraulic engineer, the surface velocities increase with the height, that is, that $r_2 > r$, and $r_3 > r_2$, it follows that the amount of the sinking of the water surface at higher stages must be in the inverse ratio of the product $b_2\,r_2$ to $b\,r$ and $b_3\,r_3$ to $b_2\,r_2$, and must therefore become smaller and smaller as the river rises, that is, $h_2 < h$ and $h_3 < h_2$.

It is evident from the above that the following criteria must be true in regard to decreases in the height of the water stages, i. e., in the sinkings of the water surface which result from a permanent deepening of the river bed.

1. The decrease in the height of the water must be greatest in low stages,

and must be proportionately less in higher stages, and must reach its minimum in floods.

2. The decrease in the height of the water which results from the deepening of the river bed must be the same, for the same average monthly stage, because this deepening only takes place slowly and gradually during a long period of years, and the stream may therefore be considered as unchanged during the period of a year.

3. If the gauge readings of a long river, taken in the same period of time, are compared, the amount of sinking of the surface, due to works of improvement or deepening of the bed, will never be the same at the different stations, but on the contrary vary greatly from each other, because on a long river only some especially bad stretches are improved gradually and in different periods of time, and because many stretches which are in a better condition, and especially those inclosed in rocky banks, are left unchanged, and because it frequently happens that the various stretches are improved according to different systems, which as the nature of the evil varies may have quite different effects upon the stream; and finally, because in the lower unimproved portion of the river, bars, sand banks, and raising of the bed occur, and therefore it happens that an equal amount of the deepening of the bed and sinking of the water surface resulting therefrom throughout the whole river is impossible.

4. When a river is radically improved, and particularly when it is considerably shortened by means of cut-offs, its bed will not only be deepened, but its slope and the velocity of its current will be increased, and as a result of this its water surface will be lowered, and sometimes even more than by deepening its bed.

But these sinkings of the water surface must possess these criteria: that they are equal at the same station, during the same months, at the same stages, independent of the years, and that even in the same improved portions of a long river they will vary with the extent and manner of the improvement; and finally, that a sinking in the water surface cannot take place on those portions of the river near its head or mouth which are not affected by the works of improvement.

In order to furnish the proof that the considerable decrease in the heights of the stages of European rivers, *i. e.*, the sinking of their water surface, as shown by me as well as by Grebenau, director of hydraulic works, to exist, did not result in consequence of the power of erosion of the water nor from the partial improvements made of these rivers at different periods of time, but is solely due to a decrease in their discharge, I procured authentic gauge readings at numerous stations on the Rhine, Danube, and Elbe, divided them into two equal periods of observation, calculated for each the monthly and annual mean stage, as well as the highest and lowest stages which occurred, and plotted them on sheets 1 to 6, in order that they might be more readily seen and compared. These sheets clearly exhibit the observations taken during a long period of years, and from them the following conclusions can justly be drawn, viz:

(*a*.) The decrease in the height of water stages, *i. e.*, sinking of the water surface of the Rhine and Elbe, is in direct opposition to the criterion laid down in my preceding reply to A under the head of 1; that is, it is the least at low and increases considerably at medium and high stages, and can, therefore, not be the result of a deepening of their beds.

The decrease shown at the six stations on the Danube did not agree with the above, because, as will hereafter be shown, the stations at Dillingen, Linz, Vienna, and Pesth were changed during the period of observation.

(*b.*) The decrease in the height of water stages at the same stations, at the same stages of the river and during the same months varies considerably, and the amount thereof depends far more upon the season of the year than upon the mean stage of the rivers, and this observed result further proves that the sinking is not due either to the deepening of their beds or to their works of improvement.

(*c.*) It will be further seen from these plottings that at several stations on the same river, even when they are separated by great distances, non-improved portions and stretches flowing over rock, the decreases in the height of the water surface, for the separate months, are the same or very nearly so, which does not agree with the criteria laid down in Nos. 3 and 4 of my reply to A.

RESULTS OF SPECIAL OBSERVATIONS ON THE RHINE.

It will be seen from the plotted gauge readings at the six stations on the Upper, Middle, and Lower Rhine, that the decrease in the height of the water surface is the greatest in the summer months of June, July, and August, and least in the winter months of November, December, and January, although the high water of summer which is poured down by the melting of Alpine glaciers and snow has already reached a medium stage when it arrives at Cologne and Emmerich; and, on the contrary, the winter low water of the Upper Rhine becomes high water at Emmerich by the addition of the water from its tributaries; consequently the Rhine completely changes its nature and regimen in this distance of 422¼ miles.

Mr. Von Salis, Helvetic chief inspector of public works, established the fact that the bed of the Rhine at Basle has not deepened during the relatively short period of observation from 1857 to 1872, in a communication to the Imperial Academy of Sciences at Vienna, dated April 17, 1875, from which the following is an abstract, viz:

"Gauge readings have been taken at Basle since 1808, which may be of value in discussing the question of decrease in the volume of water since the station has been fixed, and the geological condition of the river bed precludes the assumption of any deepening and in general of any change therein."

Mr. Honsell, member of the board of public works of the Grand Duchy of Baden, who is in charge of the works in the district beginning at Constance and ending at the Hessian frontier, in his letter to me dated November 30, 1877, has furnished overwhelming proof that there has been no deepening of the bed of the Rhine either at Basle or Waldshut, by pointing out the great firmness of its bed, which is composed largely of rock, and the relative level of the old bridge across the Rhine, built about the year 1225, and the streets of Basle. He concludes his letter as follows, viz:

"The assertion made by Grebenau that the bed of the Rhine is sinking or deepening must, therefore, so far as the neighborhoods of Basle and Waldshut are concerned, not only be described as improbable, but absolutely impossible."

It is certain also that there has been no deepening of the very rocky bed of the Rhine in the mountain gorge near Bingen during the short period from 1857 to 1872.

Now, if the Rhine at the four other stations shows, in the separate months during the half period of observations between 1857 and 1872,

a decrease in the heights of its water surface similar to those at Basle ;
and Bingen, it is perfectly justifiable to conclude that neither the partial improvement of the river nor a general deepening of its river bed,
but another cause which operates uniformly throughout its whole length,
has produced the decrease in the heights of the water surface at the
medium stages, and it is apparent that this cause can only be the decrease in the discharge, which varies from month to month according
to the amount of the rainfall.

In 1817 the Royal Prussian Government ordered that the most important gauge stations on the Rhine at Cologne and Emmerich be
improved, and that their zero be placed 2 feet and .62 inch below the
lowest water known at that time, and this was done. In his admirable
hydraulic work, Dr. H. Berghaus publishes tables of the gauge readings at Emmerich from 1770 to 1835, and expressly remarks that he
reduced the readings from 1770 to 1817 to the new zero of 1817, whose
height above the sea he places at 34 feet 3.44 inches.

Mr. Kluge, royal Prussian inspector of hydraulic works, at my request,
kindly furnished me with his official readings of the gauge at Emmerich from 1827 to 1873, and I was thus enabled, by comparing the tables
between 1827 to 1835, which were before me in duplicate, to substantiate the fact that the zero of the gauge had remained undisturbed
since 1817.

Since the comparison of the accurate readings of the very important
gauge at Emmerich during a period of 104 years, i. e., from 1770 to 1873,
is very interesting and quite sufficient to decide the question of the
decrease of the volume of water in the Rhine, I have plotted them on
Sheet 2, in order that they may be better comprehended.

Now, if the period from 1770 to 1835 be divided into two periods of 33
years each, and the other 38 years be considered as another period, and
the relative gauge readings in these three periods be compared, the following noteworthy phenomena will appear:

During the period from 1770 to 1820 the variations in height of the
annual high and low water, as well as mean stages, were very slight, excepting in the case of a few years. From 1820 to 1873 it will be noticed
that there were very frequent and considerable rises and falls in the
high, low, and mean stages, and more frequent changes between years
of abundant and small water supply.

The flood of 1861 rose 9.45 inches above those of 1799 and 1809, and
the low water of 1865, 1866, and 1870 fell from 3 feet 7.3 inches to 4 feet
0.42 inches lower than the lowest of the first period, that of 1802; consequently the difference between the highest and lowest stages in the latter
period has increased 4 feet 9.9 inches.

In the last period of 38 years, from 1836 to 1873, the arithmetical mean
heights of the water stages decreased in comparison with those of the
period from 1770 to 1802 as follows, viz:

The monthly stages from 10.63 inches to .... 4 feet 4.55 inches.
The annual stages... 2 feet 4.4 inches.
The highest stages .... 11.46 inches.
The lowest stages .... 2 feet 2.2 inches.

The observations just mentioned furnish perfectly explicit additional
proof that in the last decades the Rhine at Emmerich has undergone a
considerable change in its character and regimen, and from which we
may conclude that there has been a change in the rainfall in the whole
region of its headwaters. Furthermore, since there have been no extensive and radical improvements of the Rhine by cut-offs at Emmerich
since 1802, the foregoing considerable decrease in the stages of the river,

; *i. e.*, the sinking of its water surface at all stages and seasons, can only be explained by the circumstances that its discharge must have decreased considerably.

The reason why the decrease in the height of water surface proves to be greater at Emmerich than at the five upper stations is that the decrease of the volume of water in the whole upper river valley, and in all the springs, creeks, and rivers which emanate therefrom, are all concentrated there.

RESULTS OF SPECIAL OBSERVATIONS ON THE ELBE.

It will be found, from the plotted readings on Sheet 3 of the gauges on the Elbe at Dresden, Riesa, and Magdeburg, from 1837 to 1872, and those on Sheet 4, taken at Dresden during a period of 68 years, from 1806 to 1873, that almost the same circumstances exist relative to the decrease in the height of its water surface and increase in its variations as have been described in the foregoing paragraph on the Rhine, and, therefore, I will not discuss them here again.

A more comprehensive view of the decrease of water in the Elbe can be obtained from the plotted readings of the gauge at Magdeburg, taken during a period of 142 years, from 1728 to 1869, which are given in my first treatise of 1873. The data described in detail in the latter, and which were collected by the international technical commissions for making examinations of the Elbe in 1842, 1857, and 1869, are not only very interesting, but fully sufficient to enable a conclusion on the question before us, relating to the decrease in height of stages in that river and sinking of its water surface, to be reached. They establish the fact that the low-water of September, 1842, was from 3.9 to 11.4 inches lower than those marked on the rocks in the river beds at Tetschen, Pirna, and Strehla for the years 1616, 1706, 1719, 1746, 1782, 1790, 1800, 1811, and 1835, and it was, therefore, the lowest during a period of 226 years.

It will be seen from the readings collected by the subsequent commissions for making examinations of the Elbe, and which I have tabulated on Sheet 3 for all of its stations to the head of tide water at Blekede, that in 1852, 1857, 1869, and 1873 the low water of its upper portion, which is thoroughly improved to Dresden and Wittenberg, by means of training walls, fell about 10 inches below that of 1842, and that, on the contrary, in the lower river, which is only improved in stretches by quay walls, the bed was raised by sand banks and the low-water stage raised.

Mr. M. W. Schmidt, royal Saxon director of hydraulic works, published in the "Civil Engineer," in Nos. 4, 5, and 7 of volume 24, valuable information in reference to the slope and gauge readings of the Elbe, in the kingdom of Saxony, and this expert expressed the opinion, based upon accurate hydraulic data, that during the last decades the mean monthly and annual, and particularly these low-water stages of the Elbe, in summer and fall, have decreased considerably, and to the extent of 2 feet 0.4 inch at the station at Dresden, and from 1 foot 3¾ inches to 1 foot 6.1 inches at Riesa.*

Mr. Schmidt further proves that in consequence of the improvements which were made on the river, a deepening of its bed and sinking of its water surface took place, which amounted to 1 foot 2½ inches at Dresden and to 7½ inches at Riesa. Since the actual decrease in the height of

---

* The gauge stations at Meissen are not mentioned here, because the period of observation has been too short, and the effect of the improvements of the river made at that point has not yet been determined.

water surface has become 10¼ inches greater at Dresden, and from 7$\frac{7}{12}$ to 10¼ inches at Riesa, it is the opinion of Mr. Schmidt, "in view of the strange phenomenon, that some other cause, not resulting from the works of improvement, must have been in operation to produce this sinking in the mean stages of the water, and which, taking all the surrounding circumstances into consideration, can only be that the volume of water in the Elbe has diminished during the last twenty years. Attention may be called to the fact that after the most thorough researches of the imperial royal counsellor Wex on other European rivers, he announced in 1873 the same conclusions which have been arrived at by the observations of the height of the water in the Elbe."

Mr. Maass, of Magdeburg, member of the royal Prussian board of public works, who has made the most thorough study of the circumstances connected with the height of the water and drifting of the ice of the Elbe, and has published the result thereof, also gives, in his letter to me dated January 31, 1878, it as his opinion that the decrease in the height, *i. e.,* the sinking of the water surface of the Elbe, is due partially to its improvement and partially to the decrease in its volume of water which has taken place.

Engineer Urbata, in "Stummer's Engineer," of 1875, ascribes as a reason for the decrease of the volume of water in the Elbe, which has been universally observed, not only the clearing of the forests, but also the total drainage of ponds. Emperor Charles IV ordered the construction of ponds in the kingdom of Bohemia at the public expense, and at the end of the sixteenth century the total area of these amounted to 540½ square miles, of which it is said only about to the extent of 58 square miles now exist.

RESULTS OF SPECIAL OBSERVATIONS ON THE DANUBE.

In making the comparison between the mean of the monthly and annual readings at the most important stations on the Danube, which are plotted on Sheet 5, I could not choose periods of observation of equal lengths, because unfortunately some are quite short, and because the cross-section and the discharge of the Danube have materially changed at several stations during the last decades. Correct conclusions from the comparison of the mean-water stages can, therefore, only be drawn by considering them in connection with the changes that have taken place in the cross-section of the river. I have therefore obtained the oldest as well as the latest cross-sections as determined by experts in the public service at the four stations at Linz, Stein, Pesth, and Old Orsova, have accurately plotted them on Sheet 6, and will now briefly discuss the results of the comparison of the gauge readings at the separate stations; at the same time, take the changes in the cross-section of discharge into consideration.

The Danube at Dillingen.

Mr. Bernhard V. Herrmann, of Munich, chief director of the royal Bavarian board of public works, in a letter to me dated December 3, 1877, was so kind as to communicate to me that in his opinion the quite considerable decrease in the height of all stages, *i. e.,* in the sinking of the water surface of the Danube, at Dillingen, in Bavaria, which was observed during the second half of the period comprised between 1835 and 1874, is due to the improvement of the Danube, which has been energetically prosecuted during the last decades.

The plot of the gauge readings at Dillingen is worthy of notice, inasmuch as the following conclusions can be drawn therefrom, viz:

The slight variations between the monthly mean, as well as the small increase in height of high-water and slight decrease in height of low-water stages, indicate that large tracts of forests still exist in the region of the headwaters of the Danube, and regulate the flow of its water at Dillingen.

The fact that the variations in the monthly mean as well as in high and low water stages are so uniform, observed during the second half of the period included between 1855 and 1874, proves that the extensive river improvement at Dillingen has not only deepened its bed, but has also increased the velocity of its current considerably.

The plot of the gauge readings at Dillingen may therefore be considered as a prototype of the relative water stages of such rivers whose discharges are regulated by large forests existing in their valleys, and also for such as have been extensively and radically improved, and particularly of those in which the slope has been increased by diminishing their lengths by cut-offs. Now, as none of my plots of the gauge readings on many other streams and at various stations resemble this prototype at Dillingen, as far as regards the very uniform stages of water and their nearly equal variation at all seasons of the year, it follows that the causes which are at work at Dillingen either do not exist at all at the other stations and on the other rivers, or exist in only a very slight degree.

The Danube at Linz.

The plot of the gauge readings at Linz, taken during the second half of the period included between 1849 and 1875, shows scarcely any variation in the months of April, May, June, July, and August, which are the five richest in water supply; but, on the contrary, during the other seven months, shows a decrease in the height of water stages or sinking of the water surface of from 1.42 to 8.32 inches, of 4.8 inches in the annual mean, of 7.22 inches in high and 0.5 inch in low stages of the river.

It will be seen, however, from the cross-sections of the river, accurately taken in 1850 and 1877, only 54 feet 8 inches above the gauge at the bridge, and which are plotted on Sheet 6, that the cross-section of discharge was diminished by the construction of terraces and revetment of the river banks about 452 at the zero, about 646 at the medium, and about 1,604 square feet at the 19-foot 8-inch stage; and that in consequence of this decreased cross-section the surface of the water during the period included between 1850 and 1875 must have been raised as follows, viz: At the zero stage, about 6.66 inches; at the medium stage, about 9.84 inches; and at the high stage, about 1 foot 10.41 inches.

If, therefore, the cross-section of discharge had not been diminished, the sinking in the water surface of the river during the second half of the period included between 1849 and 1875 would have amounted as follows, viz: At the zero stage, to about 7.8 inches; at the medium stage, to about 1 foot 1.12 inches; and at the high stage, to about 2 feet 4.44 inches.

This can only be the result of a decrease in the discharge of the Danube at Linz, since it has not been improved at that point, and there has been no deepening of its bed.

The Danube at Stein.

The gauge readings of the Danube at the city of Stein, near Krems, are of the highest importance in this discussion, because at that place one of the oldest wooden pile bridges spans the Danube, and there is a wharf which is very much used. In consequence of the frequent renewal of the piles of this bridge, and the examinations made to determine the best location for its draw, frequent accurate cross-sections of the river were taken, and the gauge readings since 1829 were also preserved. I found, in prosecuting my researches, that no improvement of the Danube was made after 1829 until 1874, when its bed was narrowed at the bridge; and, therefore, I have considered neither the alterations in the water stages nor changes in the bed of the river caused thereby since 1874.

I also obtained the official cross-sections of the stream taken at the bridge at Stein in 1838, 1848, 1858, 1869, and 1872, compared them, and finally plotted the first and last ones together on Sheet 6. From these it will be seen that the bottom of the river-bed,* which is here composed of gravel, was, it is true, at times deepened by high waters, but then again filled up by them, so that the mean depth of the cross-section taken in 1872 differs only .2 of an inch from that of the one taken in 1838. It follows from this that a general sinking of the bottom of the river bed was not effected by the power of erosion of the running water, and that, therefore, Grebenau's theory seems unfounded.

Now, it will be perceived by examining the plotted gauge readings at Stein on Sheet 5, that in the second half of the period included between 1853 and 1873, the heights of all stages of the river have decreased, as follows:

The mean stages by from 3.08 inches to 1 foot 4.58 inches; the annual mean stages by 8.58 inches; the mean high stages by 1 foot 10.76 inches; the mean low stages by 8.47 inches.

Now, as the Danube was not improved at Stein, nor any deepening in the bottom of its bed took place, the assertion that this decrease in its water stages is alone due to the decrease in its discharge can be made with the fullest justification.

The Danube at Vienna.

The changes which have taken place in the cross-section of discharge of the Danube at Vienna could not be determined, because it was divided at this place into two large arms and the Vienna Danube Canal, which is 54 yards and 2 feet wide, and because a number of smaller subsidiary channels which existed just below the city were closed. It can, however, be generally asserted that the cross-section of discharge decreased considerably during the period included between 1851 to 1874, by reason of the fact that the southern arm, the so-called "Kaiserwasser,"† 162 yards wide, which formerly was 6 feet 6¾ inches deeper than the zero, and which had a considerable discharge, is so filled up that now, at the zero stage, there is none, while at the same time the cross-section of discharge of the northern arm has changed but little.

If, notwithstanding this, the tabulated comparison of the water stages and the plot thereof on Sheet 5 show that the monthly and annual mean as well as the high and low water stages for the period included between 1851 to 1874 have decreased from 5.1 inches to 1 foot 8.9 inches,

---

* The original supposition of the government experts, that a portion of the river-bed was rocky, was subsequently proven to be incorrect by the deepening which took place; and the rocky ledge, which shows itself in the river-bed about 1,968 yards below, either does not reach up to the bridge, or, if it does, lies at a great depth.
† Imperial stream.

as they do, it follows that the discharge of the Danube has decreased during that period.

The gauge readings since 1874 are not considered in this discussion, because the cut-off at Vienna, which has the normal breadth and depth of the stream and a length of 4 miles and 13 yards, was opened in 1875, the old channel completely closed, and the gauge station transferred to the cut-off, and therefore no reliable conclusions could be drawn by comparing the readings before and after the opening of the cut-off.

The Danube at Old Orsova.

The gauge readings of the Danube at the city of Old Orsova are very important and very satisfactory in the discussion and decision of this question of the decrease in the volume of water, because all the larger tributaries empty into the stream above this place, and, being in consequence thereof a mighty stream, replete with water, the abnormal stages of water in single tributaries, caused by particular elementary actions in their valleys, have no influence on its discharge at Orsova.

Again, in consequence of the great difference in the configuration, composition of the soil, and climatic conditions of the several valleys of the tributaries which flow into it from the south, west, and north, these carry off their high and low water, at different times, the Danube at Orsova may be considered as a great recipient and regulator of the high, medium, and low waters of its tributaries.

For these reasons, the most reliable conclusions relative to the discharge of the stream produced by the water which it receives from its whole upper valley can be drawn from the gauge readings at this station, but only then if it is proven that during the period in which the readings were taken there has been no change in the direction of its current, its cross-section of discharge, or in its slopes.

It can be seen from any large river map of Europe, and from works published on hydraulic subjects, that the bed of the Danube for a distance of 65.86 miles below Orsova is situated in a vast mountain gorge of the southern chain of the Carpathian ranges, and that throughout this whole distance its sides, and throughout most of it its bottom, consists of rock, from which it may safely be inferred that the direction of its current and its cross-section have remained unchanged during centuries.

It will further be seen from the plan and cross-sections made by the very intelligent Hungarian hydraulic engineer, Paul Vásárhelyi, between 1832 and 1834, which I published, together with a lecture entitled "Ueber die Schiffbarmachung der Donau am Eisernen Thore und an den sieben Felsenbänken oberhalb Orsova,"* in the "Zeitschrift des österreichischen Ingenieur und Arcaitekten Vereins,"† in No. 10 of 1872, that at a point 961.85 yards above Orsova the Danube had a nominal width at the zero stage of $462\frac{1}{2}$ yards and a depth of 38 feet $10\frac{1}{2}$ inches in the center of its channel.

Now, since Sir John Wawra, imperial and royal director in chief of the board of public works, who was sent to Orsova as a member of the international commission of experts to devise plans for the improvement of the Danube, took its cross-section in 1873 in the most careful manner, which cross-section is plotted on Sheet 6, and found that the river, at a point 961.85 yards above Orsova, was $463\frac{1}{2}$ yards wide and 38 feet

---

* On making the Danube navigable at the Iron Gate and at the seven rocky ledges above Orsova.
† Journal of the Society of Austrian Engineers and Architects.

5½ inches deep in the center of its channel, the proof is furnished by the approximate coincidence of these measurements that at Orsova the bed of the river, its depth, and its cross-section of discharge have remained unchanged since 1832, which is quite natural, since its cross-sections are bounded by rocks, both above and below that city, and make a change therein impossible.

Although it is true that the first private "Danube Steam Navigation Company" between 1847–1849, and subsequently in 1855, the imperial royal Austrian Government did remove, by blasting, some projecting rocks which endangered navigation at the lower end of the falls at the Iron Gate, yet these could not have the least influence upon the stages of the river at Orsova, because the sections of the removed rocks were insignificant compared to the cross-section of the river, and because they were a part of a ledge which is situated 5.9 miles below that city at the foot of the falls, consequently below its crest, and whose surface is about 23 feet below the zero of the gauge at Orsova.

Now, as it can be seen from the profile of the river, which I published in 1872, that, between Orsova and the Iron Gate and at the distances of 2.8 and 4.04 miles, two continuous ledges of rock, whose upper surfaces lie only 4 feet 4 inches and 4 feet 8 inches below zero, stretch entirely across the river bed, it follows further that it would be a downright impossibility for the slope of the Danube to have changed at that point, and it can therefore be safely asserted that, during the period from 1840 to 1875, in which the gauge readings were taken, the direction of the current, cross-section, and slope of the Danube remained unchanged at Orsova.

A comparison of the tabulated readings of this gauge, taken during the second half of the period of observation between 1858 and 1875, and the plot thereof, as presented on Sheet 5, will show that the mean of the monthly and annual as well as the high and low water stage has decreased from 6.18 feet to 2 feet 5 inches, and this general decrease in the volume of water, i. e., in the height of its water surface at Orsova, can only be accounted for by the fact that the discharge of the stream has diminished there.

It is notorious that, during the last decades, in several of the tributaries of the Danube—as, for instance, the Theiss and the Save—the floods sometimes rise higher than they formerly did, and their discharge is thereby increased, and that their floods are frequently poured into the great bed of the Danube and pass Orsova simultaneously with the water from other tributaries which are at a medium or low stage, yet the gauge readings, taken during the second half of the period of observation, show that these accumulated waters no longer reach the same height which they formerly did, and that the heights of the greatest floods are not as great as they formerly were by 1 foot 9.65 inches. The facts thus furnished by the large masses of water, collected together by nature itself in the Danube at Orsova, certainly will furnish every one with clear and indisputable proof that the high floods which sometimes occur in several of the tributaries do not compensate for the continued decrease in the discharge of the others at their low and medium stages.

By comparing the readings of the six gauges on the Danube it will be seen that at Stein, Vienna, and Orsova the decrease in the height of the water stages is greatest during the month of October; consequently, a uniform cause throughout the whole valley must have operated to produce the uniform effect, and that cause must have been a decrease in the rainfall.

It will also be seen by this comparison that the regimen of the Danube is partially changed in its course from Linz to Orsova, as the greatest discharges at Linz, Stein, and Vienna take place during the months of June, July, and August, i. e., during the period when the snow on the Alps melts, and on the contrary at Orsova during the months of April and May, i. e., during the rainy season of spring.

Although the foregoing proof, furnished by the decrease in the discharge throughout the whole length of the three principal rivers of Central Europe, i. e., the Danube, Rhine, and Elbe, of the correctness of my theory in regard to the decrease in the volume of water in springs and rivers would seem to suffice, I have, nevertheless, in the interest of science and for the purpose of inciting the arrangement and publication of further observations pertinent to this subject, and finally, in order to make the thorough study of this highly important hydraulic question by my fellow professionals and naturalists more easy, at the end of this treatise added a tabular exhibit compiled from the gauge readings collected by me and Mr. Grebenau at 51 stations on 13 rivers. These readings were divided into periods, and from these the annual mean height of the water stage, as well as of the high and low stages, were calculated, and finally from these ascertained the decrease in the readings during the second half of the period of observation, in order that the latter at separate stations and rivers could be more easily compared with each other.

I take the liberty of giving the following explanation of the manner in which the calculations for this exhibit were made. I not only calculated the monthly mean stages because a comparison of the latter shows not only the changes which took place in the regimen of the rivers, but also clearly the fact that the decrease in the height of their water surfaces could not be alone explained by a deepening of their beds.

In this tabular exhibit of gauge readings I took the mean of the various high and low water stages which occurred in each year, but even these do not furnish so clear and instructive a picture of the relative discharge of a river during long periods as the plots of its lowest, medium, and highest stages which I presented in my first treatise for the Rhine, Danube, Elbe, Vistula, and Oder, and in this on Sheets 2 and 4 for the Rhine at Emmerich and the Elbe at Dresden. By an examination of the annexed tabular exhibit and the calculated means of the gauge readings at 51 stations on 13 rivers, the following important conclusions may be drawn, viz:

1. Of the 160 differences in the gauge readings which are given in the table, 149 show a decrease and only 11 an increase in the height of water stages, and it can easily be proven that the latter were caused either by the present increase in the height of flood stages or by a decrease in the cross-section of discharge or by the bottom of the bed of the stream being covered with sand.

2. The average annual decrease in the height of the water surface, i. e., sinking of the water surface, is not only different on different rivers, but also for the different stations on the same river, and this is also the case with the low, medium, and high water stages, and evidently result from a difference in the amount of the rainfall in the different parts of the valleys. Not only the decrease in the water supply caused thereby, but also the deepening of the river bed resulting from works of improvement at certain localities, and finally, also, the configuration of the cross-section of discharge must have exerted a great influence in lowering the water surfaces. For this reason Grebenau's attempted calculations, based upon an equal mean in the decrease in the height of the water surfaces for all of the 14 rivers and 75 gauge stations discussed by him, as well

as for the different stages thereof, are entirely inadmissible, since only incorrect results can be obtained from them.

3. Since the closest examination of the plot of the gauge readings on the five rivers, Rhine, Danube, Elbe, Vistula, and Oder, presented in my first, as well as in this treatise, does not reveal that in the case of these rivers the number of the years which are rich or poor in water supply is not fixed in duration, or that they change regularly, or finally that they return at fixed periods, it cannot be determined during how many years gauge readings must be taken in order to be able to calculate from them a thoroughly reliable mean, and consequently cannot be determined, either theoretically or by experience, how long it is absolutely necessary that the two periods of observation which are to be compared should be. But it is clear to all that the longer the periods of gauge readings are, the more correct and reliable will be the mean calculated from them, since the years which are either very rich or poor in water supply, and which undoubtedly occur in the one or the other of these periods, have so much the less influence upon the calculated mean.

In consideration of the foregoing, the calculated decrease in gauge readings given in the annexed exhibit must have a greater value than such as were deduced from readings made during a shorter period.

It is also clear from this exhibit that the mean decrease in the height of the water derived from readings taken during a shorter period, and that the sinking of the water surfaces has become considerably greater in the last decades than formerly, which is explained by the fact that in the last decades there have been a greater amount of clearing, drainage of ponds and marshes, and improvement and irrigation of large tracts.

No.	Names of rivers and gauge stations.	Sinking of the annual mean of the gauge readings
		In inches.
	I.—RHINE.	
1	Basle	0. 114
21	Bingen	0 24
25	Emmerich	0. 40
	II.—DANUBE.	
31	Stein	0. 41
32	Vienna	*0. 425
34	Old Orsova	0. 9
	III.—ELBE.	
35	Dresden	*0. 197
40	Magdeburg	0. 394
	IV.—VISTULA.	
42	Cracow	0. 433
43	Kurzebrack	0. 558
	V.—ODER.	
44	Küstrin	0. 114
	X.—SEINE.	
53	Paris	†0. 59
	XI.—GLOMMEN.	
54	Nastangen	0. 11
55	Sarpfos	0. 24
	XII.—MISSISSIPPI.	
56	Natchez	0. 697

* The amount of the sinking of the water surface due to the deepening of the bed by improvement of the rivers was deducted at Vienna and Dresden.
† At Paris the mean of the decrease in gauge readings at the highest and lowest stages were taken.

3 W

4. From the annexed tabular exhibit the foregoing schedule is obtained by considering only the variations of the mean stages which are really the most noteworthy, at the most important stations at which material changes have taken place in the bed of the stream, and finally those that are derived from readings made during the longest periods.

From the schedule it will be seen that the decrease in the gauge readings increases in different rivers with the greater extent of their valleys, which seems to me to be another proof that the decrease in their discharge is due to a decrease in rainfall and the consequent decrease in the volume of water.

Reply to B.—The supposition of some of my opponents that it cannot be reliably concluded from a decrease in the gauge readings on a river that its discharge has decreased is unfounded, as the following proofs will show:

Mr. Grebenau, who was known as one of the most skillful hydraulic experts in the measurement of discharge of rivers and streams in Germany, while stationed at Germersheim in his former rank of inspector of public works, measured the cross-section, the slope, and the velocity of the current of the Rhine thirty-eight times at six points near Neuburg, Pforz, and Germersheim, and with all possible accuracy, and from these he calculated the quantity of water which the Rhine had discharged during the period of twenty-eight years included, between 1840 and 1867, which calculations I published in my first treatise of 1873. From these measurements Mr. Grebenau calculated the mean discharge of the Rhine, during the period from 1840 to 1867, to have been 41,596 cubic feet per second, and that the reading of the gauge at Sonderheim, corresponding to this discharge, should be + 3'3.3685". The arithmetical mean of all of the readings taken at this gauge during the above period was found to be + 3' 1", so that the difference between the two amounted to only 2.3685".

Mr. Grebenau, in communicating the result of these measurements to me in his letter dated February 2, 1872, gives the following opinion, viz:

" From this it follows that the mean stage of the river nearly corresponds to the arithmetical mean of the gauge readings, and that the error would have been small if in the beginning the discharge of the Rhine had been deduced from this arithmetical mean. This is a new and not unimportant law, which saves great and extensive calculations."

Grebenau, in his pamphlet on the results of gauge readings on the Rhine and Mosel, published in 1874, on page 20 makes the following remark bearing upon this point :

"Although the annual mean discharge of the Rhine deduced from the mean stage is nearly in the same ratio as the mean of the gauge readings, yet it does not follow from the difference in these mean stages, which was observed during the two periods, before and after 1840, that its discharge has decreased, since the difference is solely the result of the improvement of the river."

That the mean discharge of other rivers can be nearly accurately calculated from the mean of the gauge readings is apparent from the following results of measurements :

Mr. Harlacher, professor of the science of engineering at the German Polytechnical Academy, at Prague, measured with the greatest accuracy the discharge of the Elbe at Herrnskretschen, on the boundary line between Bohemia and Saxony, at the same cross-section, and at its various stages, and calculated from these the discharge during the twelve months, from July 1, 1871, to the end of June, 1872, and although during the period there were five floods (and among them the extraordinary one of May, 1872, caused by bursting of clouds), nevertheless, Mr. Harlacher found that the difference between the mean of the gauge readings and the height of the water surface due to the mean discharge of the river, during these twelve months, amounted to only 3.15 inches, and after a more accurate calculation to only 1.575 inches. (See "Beiträge zur Hydrographie Bohmens," von A. R. Harlacher, Prag, 1872 and 1875.[*])

It should here be remarked that in both cases calculated by Grebenau and Harlacher the mean of the gauge readings was smaller than the height of the stage corresponding to the mean discharge; furthermore, that the difference amounted to nothing in months during which only moderately high waters occurred, as will be seen from Harlacher's calculations, and on the contrary in months during which very high waters occurred the difference becomes somewhat greater, but yet on an average, during long periods, it will be found very small.

Now, as it may be assumed that the conditions just mentioned existed in the other streams during the various periods of gauge readings, and in further consideration of the fact that the height of the water stage corresponding to the mean discharge can only be determined by very difficult and extensive hydraulic measurements and calculations, in which considerable mistakes may easily be made, every experienced hydraulic engineer will agree with me that in such stretches of streams, in which the cross-section and the slope have remained unchanged, it may be concluded with complete justification and reliability, from a decrease in the mean of gauge readings taken during a long period, that their discharge has decreased.

Reply to C.—The supposition expressed by some of my opponents—that the discharge of rivers and streams has probably not decreased but that only their regimen has changed, since although at present that discharge at low and medium stages is smaller, yet it is very much greater at high stages, and that therefore the decrease in the former cases may be compensated by increase in the latter cases—is plainly incorrect, since on several rivers and at many gauge stations thereon the readings of high stages have decreased during the second period of observation. This has taken place on the Rhine at Basle, Worms, Bingen, and Cologne; on the Danube at Stein and Orsova; on the Vistula at Cracow; and on the Oder at Küstin. The supposition is further increased because in calculating the annual arithmetical mean of the gauge readings, the high-water stages are included, and because the monthly and annual mean of high-water stages have been found to decrease at almost all of the 51 stations on 13 rivers, which are compared in the tabular exhibit, and finally, because, if only a change in the regimen had taken place, the discharge during some months should decrease, and in the other months increase, a circumstance which has not taken place on any of the streams which are mentioned, except at a few gauge stations.

In my description of the circumstances attending the discharge of the Rhine at Emmerich and the Danube at Old Orsova I have already given the most striking proof that the decrease in low and medium stages is not compensated by that of the high stages which occasionally occur.

Reply to D.—The assertion made by Mr. Sasse, member of the board of public works, in his treatise upon the relations of the Elbe at Torgau,[†]

---

[*] Contributions to the hydrography of Bohemia, by A. R. Harlacher, Prague, 1872 and 1875.

[†] Zeitschrift für Bauwesen von Erbkam. Jahrgang, 1874. Journal of Architecture, by Erbkam, 1874.

that no reliable conclusions can be reached that the discharge of streams has decreased until gauge readings have been taken during a period of at least 200 years could only be true if it were established that the years which are either very rich or very poor in water supply alternate and return at long intervals of about 40 or 50 years, for then it would be necessary at any rate to take at least two periods rich and two periods poor in water supply in order to obtain a reliable mean. But since it is apparent from the plotted gauge readings, in this as well as in my first treatise, of five rivers, during periods of from 60 to 142 years, that the periods rich or poor in water supply do not habitually last longer than from three to five years and that they do not recur in regular order, and since the further assumption of Mr. Sasse that the discharge of streams during years rich is frequently three times as great as that during years poor in water supply is true only of the smallest number of streams and only very seldom, of which my plotted gauge readings will convince any one, his assertion that in order to reach a reliable conclusion as to the decrease in the discharge of rivers gauge readings must be taken during a period of at least 200 years appears unfounded, since from my plotted gauge readings it appears that we can—if we take these during periods of from 40 to 60 years, divide them into two periods, calculate the mean of each of these, and then compare these means—infer from an increase or decrease in these that the discharge has become greater or less.

I should remark, however, that if in any year in either of the half periods of observation a flood, produced by extraordinary action of the elements, should occur, as happens rarely during 100 years, it would certainly seem advisable in order to obtain a correct mean of the gauge readings to exclude such an abnormal year and substitute for it the year of the other half period which was richest in water supply. I did not, however, deem it necessary to make the correction just referred to in my exhibit and plot of the gauge readings.

Reply to E.—The opinion of some of my opponents that a decrease in the discharge can only be reliably established by making direct measurements of it on rivers and streams from time to time during long periods, seems at first to be evident, but if we calmly consider the whole proceeding and the manner in which such measurements and calculations must be made, the surprising result will be reached that to establish the decrease in discharge by direct measurements is quite impracticable and almost impossible, as I now propose to show.

First of all, every hydraulic engineer will agree with me that it is an utter impossibility to make during 30 or 40 years uninterrupted measures of discharge, and at the varying stages at such stretches of streams in which the current, the cross-section, or the slope are constantly changing, and then by comparing the volumes of discharge of the two periods to determine whether an increase or decrease has taken place. In this matter every experienced hydraulic engineer will agree with me that the discharge can only be measured on those stretches of the stream in which the current, cross-section, and slope have remained unchanged during 30 or 40 years, as I have shown to be the case on the Rhine at Basle and on the Danube at Old Orsova.

Now, if we assume that, in order to decide the question whether the discharge of a stream has increased or diminished, a hydraulic engineer is intrusted with measuring it on an unchanging stretch of it during a period of 30 years, and assuming also, for the sake of simplicity, that in this stretch only those stages occur—that is, the lowest h, the medium h^1, and the high h^2—and that he has in the first year executed his trust,

and has found by accurate measurements that at the selected cross-section
of the stream the discharge per second at the stage *h* was M, at the stage
h¹ was M¹, and at the stage *h²* was M².

Now, if the engineer, after 5, 10, 20, or 30 years have elapsed, makes
similar accurate measurements of discharge at the same cross-section at
the same three different stages, it is evident that, if the current, cross-
section, and slope of the stream have remained entirely unchanged, he
will find exactly the same quantities of discharge, M, M¹, and M², and
it will be impossible, in spite of these measurements, after a lapse of 30
years, to determine whether there has been an increase or decrease in
the volume of discharge, since the gross amount of the latter depends
upon the duration of the low, medium, and high water stages.

What has just been said concerning the three stages of the water is
also true of all the variable stages of a stream during the whole period
of observation, and therefore my assertion that the volume of discharge
of a stream cannot be determined by measurements at an invariable
cross-section, when the discharge will during the whole period of observation be the same for the same stage of water, but only from the calculated mean of the gauge readings whether the volume of discharge
has increased or decreased, seems well founded.

If it is desired, however, to calculate and compare with each other the
gross volume of discharge which takes place during two periods of
about fifteen or twenty years each, the following method, already
adopted by Mr. Grebenau in calculating the gross volume of discharge
of the Rhine at Sondernheim from 1840 to 1867, and by Mr. Sasse,
member of the board of public works in his calculation of the gross
volume of discharge of the Elbe at Torgau * from 1831 to 1850, must be
pursued.

First of all the curve must be constructed based upon accurate measurements at various stages, and at a constant cross-section, from which
the discharge at any stage can readily be obtained for a second of time.
Then from the table of gauge readings the exact number of days of the
period during which each stage lasted, and then multiplying the discharge M M₁M₂, &c., per second at each stage by 60″ × 60′ × 24 =
86,400 seconds, and this product by the number of days of each half
period of fifteen or twenty years, during which the stage lasted, and add-
ing together all of these sums for each of the two periods, we can learn
from the resulting gross amounts whether in the second half of the period of observation the discharge has decreased.

Now every experienced hydraulic engineer must agree with me that
it is very difficult to measure the discharge of a large stream at the different and especially at the high stages, also to obtain the correct curve
from these, and also to make the weary, time-robbing, and tiresome collection of the same stages from the tables of gauge readings taken during a long period, and to make the unavoidable rounding off of those
that are nearly alike to get them into designated groups, and finally to
perform the similar time-consuming multiplication and addition of colossal sums, in consequence of which a very busy engineer must intrust
the labor to an assistant, and that mistakes are very easily made which
may be greater than the actual difference between the gross amounts of
discharge of the two half periods of observation, and actually reverse
the result, and that on the contrary the calculation of the annual mean
stage of the river, as well as the mean of the gauge readings taken dur-

* Zeitschrift für Bauwesen von Erbkam, vom Jahr 1874. Journal of Architecture,
by Erbkam, 1-74.

ing a long period of time, can be made easily and without error from the tables of gauge readings.

Under these circumstances hydraulic engineers will probably now acknowledge that the result of Grebenau's experience in the measurements and calculations of discharge so numerously made by him, that the arithmetical mean for a long period, calculated from the tables of gauge readings, is nearly identical with that stage of the river which corresponds to the mean of the discharge obtained from measuring the volume of discharge during that period, is highly important, and it is therefore easier and more reliable to obtain the mean stage of a river during a long period from the tables of the gauge readings taken during that period.

Now I will mention another simple method by which, without the very difficult and time-consuming measurements of discharge, even a non-professional can calculate the amount of decrease in the discharge of a stream during a certain period. It must be here remarked that only such stretches and such gauge stations, where the current, cross-section, and slope of the stream have remained unchanged during the whole period of observation, are suitable for reliable calculations of the actual decrease in discharge, and I will therefore, to illustrate my method by examples, take the stations on the Rhine at Basle and on the Danube at Orsova, both of which are fully suited thereto.

According to the comparison of the gauge readings in the annexed tabular exhibit in No. 1, the annual mean stage of the Rhine at Basle decreased on an average during the second half of the period of observation from 1809 to 1868 (that is, from 1839 to 1868) at the rate of .114 inch per annum; that is, an upper layer of water .114 inch thick was discharged less each year. Mr. Grebenau, it is true, did not give the mean of the annual stage from 1839 to 1868, but it will probably be very nearly the same as that for the period from 1840 to 1872 given in No. 2 as 5.8495 feet, or 5 feet 10.194 inches above zero. If now the mean surface velocity for this stage is obtained, which can easily be done by throwing a number of floats into the stream at equal distances apart, we can, by multiplying the breadth of the stream by the mean surface velocity and the product by the established sinking in the height of the water surface, i. e., .114 inch, obtain the volume of water per second by which the discharge of the Rhine at Basle is decreased. All the data necessary in order to make the calculations for this case can be obtained from the "Internationalen Rheinstrom-Messung bei Basel im November 1867."[*] It will be found from this scientific and very thorough work that at the stage 5′ 10.194″ above the zero of the water at Basle the width of the water surface is 728.31762′, and the mean velocity of the water on the surface of the stream is 7.4144′, so that the decrease in the volume of discharge will be 728.31762′ × 7.4144′ .00952=46.7971 cubic feet per second, and by further multiplication will amount to 1,475,794,557 cubic feet per annum.

Now, as the area of the whole valley of the Rhine above Basle is given in the Swiss charts at 35,906,900,000 square meters=386,468,047,300 square feet, the foregoing decrease in the discharge of the Rhine would indicate a decrease in the depth of the rainfall of $\frac{1,475,794,557}{386,468,047,300}=$ .0038′ = .0456″ per annum.

Now, if we base our calculations upon the decrease in the height of

* International measurements of the Rhine at Basle in November, 1867. These measurements of the Rhine were described and published by Grebenau. March, 1873. Lindauer's bookstore.

water-stages during the period of 16 years from 1857 to 1872, which, under the head of No. 2 in the tabular exhibit, is shown to be .77556″ = .06463′, and take the corresponding dimensions from the work above cited, *i. e.*, the stage at 5.3344′, the width at 659′, and the mean velocity at 6.56′, the decrease of the water during the period will be found to be 659′ × 6.56′ × .06463′=279.5703 cubic feet per second and 8,816,528,981 cubic feet per annum. The amount of decrease in the depth of rainfall per annum would therefore be $\frac{8,816,528,981}{386,468,047,300}$=.0228′=.2736″.

It will be shown hereafter how nearly these calculated results agree with the meteorological observations.

During the second half of the period of 36 years, *i. e.*, from 1858 to 1875, the mean stage of the Danube at Old Orsova was 8.2116′ and the average decrease of the latter .0748′, as will be seen at No. 34 of the tabular exhibit.

According to the cross-section of the Danube at Old Orsova, taken in 1873 by Sir Wawra, chief of the board of public works, and shown on Sheet 6, the width of the stream when at this stage was 1,549′, and the mean surface velocity, taken, however, at a stage only 3.28071′ above zero, was 2,687′.

It is universally known that the mean surface velocity of a stream increases with its height, and hence I ascertained by actual measurements of the velocity of the Danube at different stages between Ofen and Pesth, and at a similar cross-section at the foot of Block's hill, that when the water rises 4.92′ then the mean velocity of the surface current is increased about 1.148′. It can therefore be assumed approximately that when the Danube at Old Orsova is at its 8.2116′ stage its mean surface velocity will be 2.687′ + 1.148′=3.835′.

The decrease in the discharge for an annual diminution of .0748′ in the height of the stage would therefore be 1.549′ × 3.835′ × .0748′= 43.2189 cubic feet per second, and consequently 1,362,951,220 cubic feet per annum.

The whole valley of the Danube has an area of 14,420 geographical square miles, of which the portion below Orsova contains 4,100. The portion above this place then contains 10,320 geographical square miles =6,093,561,254,602 square feet. The decrease in rainfall is therefore found to be $\frac{1,362,951,220}{6,093,561,254,602}$=.000224′=.0027″ per annum, and therefore during the whole period of 18 years, .0486 inch. It is no doubt self-evident that this decrease in rainfall was greatest in the mountainous regions and on the contrary less on the plains.

Reply to F.—In order to remove the doubts and exceptions in regard to the correctness of the assertion made by me that the rainfall has been decreased by the destruction and devastation of extensive forests, I believe it to be necessary, in the first place, to call attention to the opinions and treatises on this subject by distinguished experts and naturalists given in detail in Chapter I, and then to communicate the following explanations and results of my observations.

It is quite natural that the meteorological observations made in England, Scotland, and at Paris, St. Petersburg, and Copenhagen, during about 100 to 190 years, did not indicate any decrease in the volume of the annual rainfall, since these countries and cities are situated in the vicinity of the sea, and receive the rain-clouds, so to speak, from first hands. Even if it should appear from the meteorological observations at the principal cities of the continent that at these no decrease of rainfall had taken place, this can be explained by the fact that during the last decades no such extensive clearings as would cause a decrease in the volume of rainfall have taken place in their vicinities.

I can give only two instances to prove that there has been a decrease in rainfall in those regions, and especially those of a mountainous character, in which large clearings were made, because, unfortunately, in former times no meteorological stations existed in these localities.

Mr. Adam Seidel, chief forester at Bodenbach, in the Erz* Mountains of Bohemia, made precise meteorological observations uninterruptedly during the period from 1828 to 1873, the results from which were revised and published by the Imperial Royal Central Bureau of Meteorology. If the observations of 1828 and 1850, which were incompletely made, are eliminated, and then divide the remainder of the time into two equal periods of 22 years each, and calculate their arithmetical means, the following results will be obtained, viz:

Meteorological observations by Chief Forester Adam Seidel, at Bodenbach; arranged by Stanislaus Kostliny.

	Annual mean.		
	1829 to 1851.	1852 to 1873.	Decrease from 1852 to 1873.
Mean temperature of the air .... degrees Fahrenheit..	47. 66	47. 22	0. 44
Pressure of vapor... inches..	0. 2092	0. 2874	0. 0118
Humidity of the air ....per cent..	84. 6	82. 1	2. 5
Number of days of rain..	158	145. 8	12. 2
Depth of rainfall ....inches..	25. 235	23. 974	1. 259

These results substantiated the fact that in the mountainous regions at Bodenbach a decrease in the pressure of the vapor, relative dampness of the air, the number of days of rain, and the depth of rainfall has taken place in the latter period of 22 years.

The annual decrease in the depth of rainfall amounted to $\frac{1.259}{22} = 0.057$ inch.

Mr. Plantamour, the director of the meteorological observatory at Geneva, also has shown that a decrease of rainfall has taken place in the Alps. He found, namely, in the latter period of 14 years, from 1861 to 1874, in comparison with the preceding period of 20 years, on the St. Bernard, at a height of 8,116.5 feet, an increase of temperature of 0.72° F., a decrease in rainfall of .008 inch, and in snowfall of about one-half, i. e., from 32.8 to 15.93 feet. For Geneva a comparison of the last 11 years with the preceding period of 35 years shows an increase in temperature of 1.134° F., and a decrease in the rainfall of 3.3 inches, and therefore an annual decrease of $\frac{3.3}{11} = .3$ inch. These changes, according to the opinion of Plantamour, are due to the contraction or decrease in the size of the glaciers which has been observed during the last 12 years.

Now, as we found in the preceding paragraph, in my reply to E, from the decrease in the height of water stages, that there was an average annual decrease in the depth of rainfall in the valley of the Rhine above Basle during the period from 1839 to 1868 of .0456″ and during the period of 16 years from 1857 to 1872 of .2736″, the similarity of these results, obtained from an entirely different source, to the decrease of rainfall in the mountainous regions at Bodenbach and Geneva is surprising.

* Ore.

Several countries in Europe have, unfortunately however, only since the last few years established meteorological stations in the forest and mountain regions, and after the lapse of several decades it is highly probable that they will show the same decrease in rainfall as at Bodenbach and Geneva.

CONCLUSION.

As I now believe that I may assume that my theory in regard to the decrease of water in springs and rivers, announced in 1873, has been established as a fixed fact by my discussion, and then by the tabular exhibit of observations at 56 stations on 13 rivers, and finally by the abundant opinions and thorough treatises of naturalists and experts given by me, I take the liberty of making the most earnest request of my readers and colleagues that each may labor in his own sphere, so that the higher governments, authorities, corporations, land-owners, and communities may finally be convinced of the numberless disadvantages and dangers our present cultivated countries are approaching, if a limit is not set to the further devastation and destruction of forests, and that it is imperatively necessary to carry out, as rapidly as possible, the precautions and measures recommended in my treatise of 1873, and which have been also warmly approved by other authorities, and thus prevent, as far as it yet lies in man's field of labor, the calamity of a still further decrease of the water in springs and rivers which threatens future generations.

Finally, I take the liberty of saying to my esteemed readers that I am ready to furnish cheerfully all the data and results which I have collected, and all the books and pamphlets which have been published bearing upon this subject, for the examination and use of all such gentlemen who intend to still more thoroughly investigate this hydraulic question.

[Note by the translator.]

In making the foregoing translation, I have attempted to make it as literal as possible, and yet make it fairly intelligible English reading.

I have used the word "rainfall" to indicate that term of the author which literally translated would be "aqueous atmospheric precipitations," and which includes rain, snow, hail, dew, &c.

I have also changed all the metric measures and centrigrades into American measures and degrees of Fahrenheit.

G. WEITZEL,
Major of Engineers,
Brevet Major-General, U. S. A.

4 W

www.ingramcontent.com/pod-product-compliance
Lightning Source LLC
Chambersburg PA
CBHW021944190326
41519CB00009B/1141